CW01221454

The Open University

S342

Science: a third level course

PHYSICAL CHEMISTRY

PRINCIPLES OF CHEMICAL CHANGE

BLOCK 7
EQUILIBRIUM ELECTROCHEMISTRY

THE S342 COURSE TEAM

CHAIR AND GENERAL EDITOR

Kiki Warr

AUTHORS

Keith Bolton (Block 8; Topic Study 3)
Angela Chapman (Block 4)
Eleanor Crabb (Block 5; Topic Study 2)
Charlie Harding (Block 6; Topic Study 2)
Clive McKee (Block 6)
Michael Mortimer (Blocks 2, 3 and 5)
Kiki Warr (Blocks 1, 4, 7 and 8; Topic Study 1)
Ruth Williams (Block 3)

Other authors whose previous S342 contribution has been of considerable value in the preparation of this Course

Lesley Smart (Block 6)
Peter Taylor (Blocks 3 and 4)
Dr J. M. West (University of Sheffield, Topic Study 3)

COURSE MANAGER

Mike Bullivant

EDITORS

Ian Nuttall
Dick Sharp

BBC

David Jackson
Ian Thomas

GRAPHIC DESIGN

Debbie Crouch (Designer)
Howard Taylor (Graphic Artist)
Andrew Whitehead (Graphic Artist)

COURSE READER

Dr Clive McKee

COURSE ASSESSOR

Professor P. G. Ashmore (original course)
Dr David Whan (revised course)

SECRETARIAL SUPPORT

Debbie Gingell (Course Secretary)
Jenny Burrage
Margaret Careford

The Open University, Walton Hall, Milton Keynes, MK7 6AA

Edited, designed and typeset by The Open University.

Printed in the United Kingdom by Henry Ling Ltd, The Dorset Press, Dorchester DT1 1HD

ISBN 0 7492 51883

This text forms part of an Open University Third Level Course. If you would like a copy of Studying with The Open University, please write to the Central Enquiry Service, PO Box 200, The Open University, Walton Hall, Milton Keynes, MK7 6YZ. If you have not enrolled on the Course and would like to buy this or other Open University material, please write to Open University Educational Enterprises Ltd, 12 Cofferidge Close, Stony Stratford, Milton Keynes, MK11 1BY, United Kingdom.

s342block7i1.2

CONTENTS

1 INTRODUCTION

So far, the main emphasis in S342 has been on reactions that are brought about via the familiar chemical or 'thermal' route. Even here, there is at first sight a bewildering variety of phenomena – ranging from the simplest test-tube reaction, to the subtle surface processes so crucial to much of the chemical industry. Nevertheless, you have seen that it is possible to identify certain general chemical principles that govern *any* reaction, no matter how complex. Broadly speaking, these principles derive from two of the most important cornerstones of physical chemistry: thermodynamics and chemical kinetics. The combination is powerful indeed – and not only in a strictly academic context. By this stage of the Course, you should be in little doubt about the far-reaching practical importance of this two-pronged approach.

For the remainder of the Course, we turn our attention to *electrochemistry*. As before, one of our aims will be to highlight the practical importance of this alternative path to chemical change. Electrochemical processes have long been in use at one or more stages in the production of many materials in everyday use – certain polymers, for instance, and a large number of metals and the objects made from them. Equally important is the role of electrochemistry in energy conversion and storage: witness the fact that, in the countries of the western world at least, we presently manufacture 4–10 batteries *per head* every year.

The scope of applied electrochemistry is very broad, so our discussion of these applications must necessarily be selective. Indeed, for much of this Block, and the next, we shall be more concerned with establishing the underlying principles of electrochemistry – that is, with exploring the thermodynamics and kinetics of electrochemical reactions in general. Once again, you will see how this dual approach can add much to our understanding of such processes, on both a macroscopic and a microscopic level.

Certainly, we shall touch on aspects of the electrochemical industry as we go along. But the principles and methods described in Blocks 7 and 8 are brought together most forcibly in the final Topic Study of the Course. Here you will examine a problem in applied electrochemistry, the importance of which needs little emphasis: metallic corrosion, and the measures taken to slow its progress, and hence mitigate its effects.

STUDY COMMENT The material in this Block falls fairly readily into three parts. The first part runs from Section 1.1 to the end of Section 3: much of the discussion is built around a series of, largely qualitative, observations drawn from simple experiments with electrochemical cells. These experiments are demonstrated in the video sequence *Electrochemical cells* (band 8 on videocassette 2), so it would be a good plan to watch this sequence as soon as possible during your study of this Block. In the second part of the Block, we take a closer look at the thermodynamics of chemical change in general (Section 4), and of electrochemical reactions in particular (Section 5). Here you will meet one of the most important results in the whole of electrochemistry – the Nernst equation. Armed with this result, we then spend the third part of the Block (Sections 6–8) examining the scope and limitations of the thermodynamic approach to electrochemical reactions.

1.1 An electrochemical reaction: some basic terminology

There is one type of electrochemical reaction that should already be fairly familiar: **electrolysis**. Important industrial examples include the massive chlor-alkali industry, and the extraction and refining of many metals. To recall the essential features, consider the electrolysis set-up shown schematically in Figure 1.

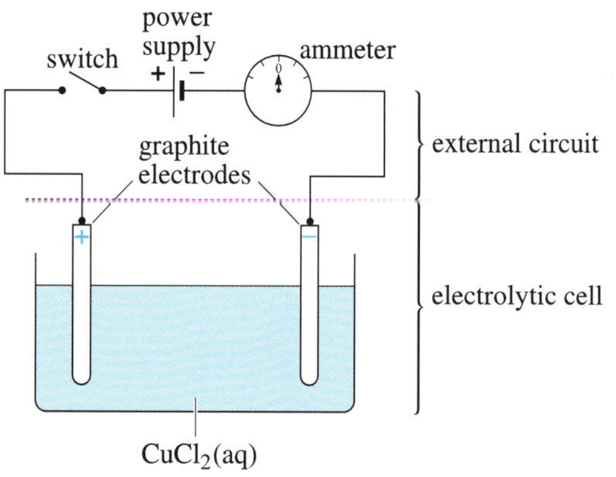

Figure 1 An example of an electrolytic cell.

In this case, the **electrolytic cell** (as it is usually called) comprises two graphite rods, or **electrodes**, dipping into an aqueous solution of copper(II) chloride ($CuCl_2$). An external electrical circuit connects the electrodes of the cell to the terminals of a suitable power supply (a battery, for example). In the terminology inherited from the Second Level Inorganic Course, the electrodes are labelled 'positive' and 'negative' according to the terminal of the power supply to which each is connected.

Now, when the circuit in Figure 1 is completed, current flows. Before long, bubbles of chlorine gas appear on the positive electrode, and a metallic deposit with the characteristic pink–brown colour of copper can be detected on the negative electrode. The interpretation of these observations is fairly straightforward.

- What ions are present in the solution?

- $Cu^{2+}(aq)$ and $Cl^{-}(aq)$. Copper(II) chloride is an electrolyte; when dissolved in water, it dissociates into ions as follows:

  $CuCl_2(s) = Cu^{2+}(aq) + 2Cl^{-}(aq)$ (1)

- Write equations for the processes at the two electrodes.

- Given the nature of the products at the electrodes, reasonable equations would be:

  at the positive electrode: $2Cl^{-}(aq) = Cl_2(g) + 2e$ (2)

  at the negative electrode: $Cu^{2+}(aq) + 2e = Cu(s)$ (3)

- In the equations above, are the chloride ions oxidized or reduced? What about the copper ions?

- Oxidation and reduction can be defined as the loss and gain of electrons, respectively. Thus the $Cl^{-}(aq)$ ions are oxidized and the $Cu^{2+}(aq)$ ions are reduced.

Is this consistent with the electrical circuit implied by Figure 1?

Yes. The negative electrode is connected to the negative terminal of the power supply. Thus electrons pour *into* this electrode from the external circuit. It therefore acts as an electron source, transferring electrons to species in the solution (Cu^{2+} ions in this instance), which are thereby reduced. For the system to continue functioning, electrons must flow *away from* the positive electrode, and hence back to the power supply. This electrode must then act as a kind of electron sink, taking electrons from species in the solution (here Cl^{-} ions), which are thus oxidized. The flow of electrons is summarized in Figure 2.

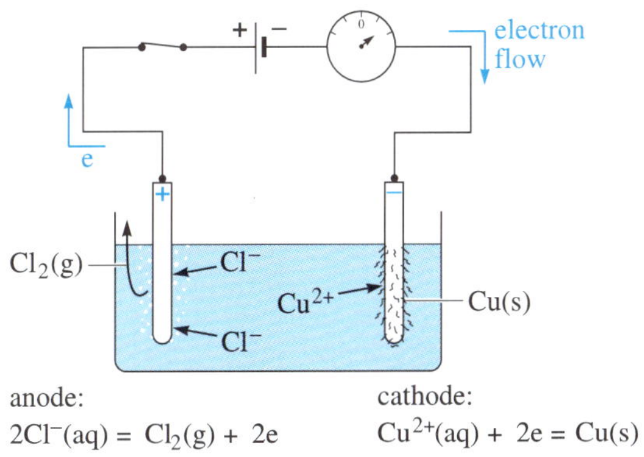

Figure 2 The electrolysis of aqueous copper(II) chloride, $CuCl_2(aq)$. Electric current is conventionally regarded as the flow of a *positively* charged substance, so the current flow is strictly opposite to the flow of electrons: this convention is very confusing. Throughout this Block, the ammeter is taken to record the direction of flow of *electrons*.

For reasons that will become clearer in the next Section, we now adopt alternative names for the two electrodes, defining them in terms of the *processes* that occur there. The definition goes as follows:

The electrode at which oxidation occurs is called the **anode**; that at which reduction occurs is called the **cathode**.

This nomenclature may be conveniently remembered by noting that **a**node and **o**xidation both begin with vowels, whereas **c**athode and **r**eduction both start with consonants. It may also help to recall that oxidation is defined as a *loss* of electrons, and that the anode is where the cell '*loses*' electrons to the external circuit.

- According to this definition, which of the electrodes in Figure 2 is the anode, and which is the cathode?

- In this *electrolytic* cell (but see Section 2.1), the negative electrode is the cathode, and the positive electrode is the anode.

Notice one further point about the system in Figure 2. For the electrolysis to work, the current must obviously be carried through the solution in the cell, as well as around the external circuit: the charge carriers in solution are ions, not electrons. Indeed, it was the conductivity of aqueous electrolyte solutions, coupled with an analysis of the substances formed at the electrodes, that provided some of the earliest evidence for the existence, and nature, of aqueous ions.

This movement of ions – or *ion transport* – is an essential prerequisite for *any* electrochemical reaction. Thus, an electrochemical system comprises a combination of electronic conductors (the external circuit and electrodes) and ionic conductor(s). It is only because of the switch over from one charge carrier to the other that the system as a whole leads to *chemical* transformations.

- With this in mind, write an equation for the *overall* reaction in the cell shown in Figure 2.

- The separate electron-transfer processes at the two electrodes must obviously balance one another; otherwise there would be a loss of overall electroneutrality within the solution. Equations 2 and 3 as written involve transfer of the same number of electrons, so the required equation is simply their sum:

$$Cu^{2+}(aq) + 2Cl^-(aq) = Cu(s) + Cl_2(g) \qquad (4)$$

Equation 4 clearly represents a purely *chemical* change. The point is made more forcibly if we include the step involved in preparing the electrolyte solution, equation 1. Thus, the net electrochemical reaction, obtained by adding together equations 1 and 4,

$$CuCl_2(s) = Cu^{2+}(aq) + 2Cl^-(aq) \qquad (1)$$

$$Cu^{2+}(aq) + 2Cl^-(aq) = Cu(s) + Cl_2(g) \qquad (4)$$

$$CuCl_2(s) = Cu(s) + Cl_2(g) \qquad (5)$$

is formally identical with the simple *chemical* decomposition of copper(II) chloride. This, then, is the sense in which electrochemistry provides an alternative route for effecting a chemical change.

But notice that there is a crucial difference between the two paths. The overall reaction in equation 5 (or equation 4) is certainly an oxidation/reduction process, but the electrochemical path exploits this fact *explicitly*. It does so by ensuring that the two complementary **electron-transfer reactions**, or **half-reactions** as they are called (equations 2 and 3), occur at *separated* (electrode) sites. As you will see time and again in this Block and the next, this separation lies at the heart of electrochemistry: it is the essence of *any* **electrochemical reaction**.

Before reading further make sure you try the following SAQ.

SAQ 1 Hydrogen is a potential fuel of the future. An obvious source of hydrogen is water, and one way of decomposing it into its elements is electrolysis:

$$H_2O(l) = H_2(g) + \tfrac{1}{2}O_2(g) \qquad (6)$$

The simple apparatus shown schematically in Figure 3 will do the trick.

(a) For the electrolysis to work, the water must be laced with a little acid. Why is this?

(b) Under the conditions implied by Figure 3, the half-reaction at the negative electrode can be written as follows:

$$2H^+(aq) + 2e = H_2(g) \qquad (7)$$

Write an equation for the half-reaction at the other electrode. [*Hint* Remember that the *overall* cell reaction (that is, the sum of the two half-reactions) must correspond to equation 6.]

(c) Identify the anode and cathode in Figure 3.

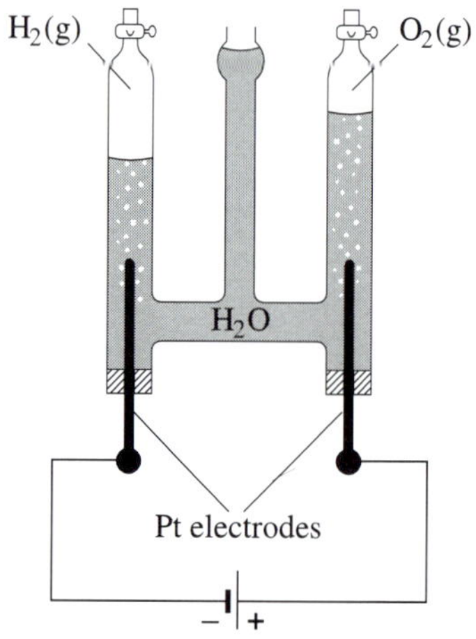

Figure 3 The electrolysis of water (small amount of acid added).

Having revised the basic terminology of electrochemistry, we now take a cue from our approach to 'ordinary' chemical reactions, and concentrate in this Block on the thermodynamics of electrochemical processes.

- Use information from the S342 *Data Book* to determine $\Delta G_m^{\ominus}$ at 298.15 K for the reaction in equation 6. What do you conclude from your answer?

- Equation 6 is the *reverse* of the formation reaction for $H_2O(l)$, so

  $\Delta G_m^{\ominus} = -\Delta G_f^{\ominus}(H_2O, l) = +237.1 \text{ kJ mol}^{-1}$

  Evidently, the decomposition of water is *thermodynamically unfavourable* at 298.15 K: it simply cannot happen of its own accord – not under ambient conditions at least.

In fact, *all* electrolyses share this feature: in each case, the underlying *chemical* reaction is a *non-spontaneous* process. (Check this by working out $\Delta G_m^{\ominus}$ for reaction 4 or 5, if you wish.) This certainly explains why the reaction must be 'driven' by connecting the cell to an external source of electricity. But is it possible to quantify this idea? Does thermodynamics provide any clues to the 'driving force' required to bring about the desired electrolysis?

It turns out that the best way to tackle this question is to concentrate first of all on a second type of electrochemical cell, where the underlying cell reaction is a *spontaneous* process. Cells like this can, and do in practice (in batteries, for example), act as producers rather than consumers of electricity. It is with these '**self-driving cells**' that the important thermodynamic factors reveal themselves most clearly.* This study, and the thermodynamic analysis that stems from it, will take us through most of the Block. Only then shall we be in a position to return to electrolytic systems, and see to what extent thermodynamics can indeed provide answers to questions like the one posed above.

* Indeed, it was through the careful study of this type of cell that 19th century scientists, such as the brilliant Walther Nernst (Figure 23, Section 5.3), first laid the foundations of much of chemical thermodynamics as we know it today.

2 SPONTANEOUS REACTIONS AND ELECTROCHEMISTRY

STUDY COMMENT As suggested earlier, this would be a good point to pause and watch band 8 on videocassette 2 – if you have not already done so. Although the text that follows can be read without this background, the simple demonstrations and animations in the video sequence should help to bring the key points 'alive'. As you work through the material in this Section, a number of important, general questions will arise. Don't worry about losing track. The questions are collected together in the summary in Section 2.5.

2.1 A spontaneous reaction in an electrochemical cell: some initial observations

As demonstrated in the video sequence, when copper metal is dropped into silver nitrate solution, the following reaction takes place:

$$Cu(s) + 2Ag^{+}(aq) = Cu^{2+}(aq) + 2Ag(s) \qquad (8)$$

It is a spontaneous process – one that can equally well be brought about in an **electrochemical cell**. A suitable set-up is shown schematically in Figure 4. Notice that it shares many features with the electrolytic systems discussed earlier: specifically, there are two electrodes (metals in this case), each of which dips into an aqueous electrolyte solution, connected together via an external electrical circuit.

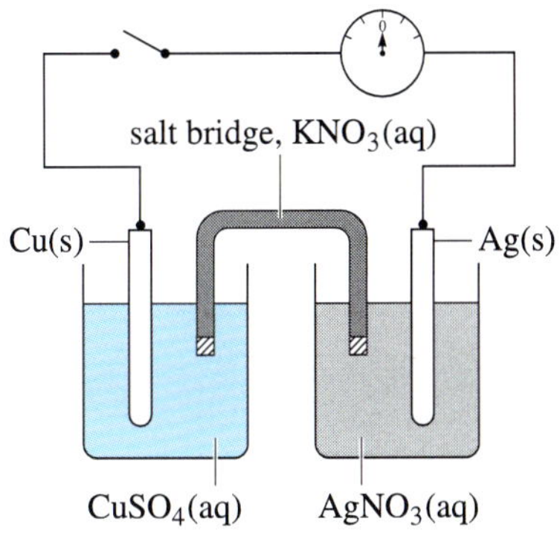

Figure 4 A typical electrochemical cell, with a salt bridge.

Closer inspection, however, reveals a number of important differences. First, the cell has been set up in two distinct parts, or **half-cells** as they are known. The reasons for this separation will become clearer in a moment. However, it immediately highlights a second characeristic feature of the arrangement in Figure 4.

What is the function of the tube labelled '**salt bridge**'?

A prerequisite for any electrochemical reaction is a pathway for ion transport *within* the cell. The electrolyte in the tube (usually a saturated solution of potassium nitrate or potassium chloride) provides such a pathway *between* the two halves of the cell. At the same time, the porous plugs in the ends of the tube prevent any significant mixing of the three solutions.

But the most important difference becomes apparent as soon as the switch indicated in Figure 4 is closed: current flows through the external circuit. In other words, the cell generates a *spontaneous* flow of electrons: it acts like a battery. Evidence that the underlying cell reaction can be represented by equation 8 comes when the switch is left closed for several hours. The result is an intensification of the characteristic blue colour of Cu^{2+}(aq) ions around the copper electrode, accompanied by the growth of beautiful needles of metallic silver on the silver electrode.

■ Which electrode is the anode and which the cathode?

■ The observations above suggest that copper atoms lose electrons and are oxidized to Cu^{2+}(aq) ions at the copper electrode, whereas Ag^{+}(aq) ions take up electrons and are reduced to silver metal at the silver electrode. According to the definitions in Section 1.1, copper is the anode and silver the cathode.

■ Write equations to represent the reactions at the electrodes.

■ anode: $Cu(s) = Cu^{2+}(aq) + 2e \qquad (9)$

cathode: $Ag^{+}(aq) + e = Ag(s) \qquad (10)$

As before, the electron-transfer processes in equations 9 and 10 are called half-reactions: they occur in the two half-cells that, together, make up the electrochemical cell. The flow of electrons is summarized in Figure 5.

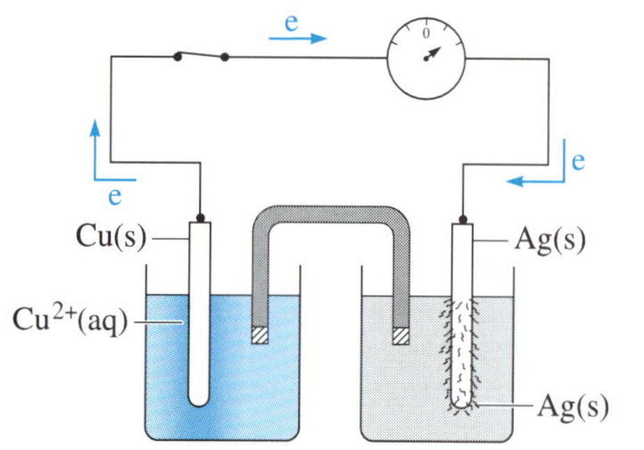

Figure 5 A spontaneous reaction in an electrochemical cell.

- If you think of the system in Figure 5 as a battery, which electrode would be the negative terminal?
- Electrons are generated at the copper anode, so this would be the negative terminal.

Notice that this sign conflicts with positive anodes (and negative cathodes) in electrolytic cells. For this reason, if no other, *you should always concentrate on the processes taking place at the electrodes: in both cases, oxidation occurs at the anode, reduction at the cathode* – see Box 1.

Box 1 Anodes and cathodes

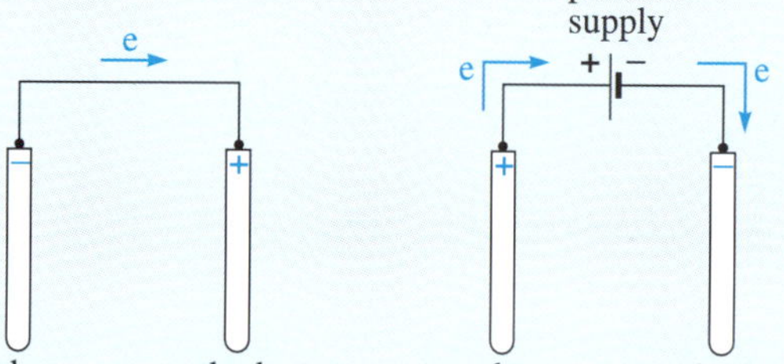

The terms anode and cathode apply equally both to 'self-driving' electrochemical cells (a), and to 'driven' electrolytic cells (b): in both cases, there is oxidation at the anode and reduction at the cathode; in both cases, electrons leave the cell via the anode and enter via the cathode. The difference in *sign* arises because of the need to drive the reaction in an electrolytic cell, by connecting the cathode to an external source of electrons – the negative terminal of a power supply.

Once again, the overall reaction is the sum of the two half-reactions in equations 9 and 10. But take care. To prevent a build-up of charge, the electrons produced at the anode must *all* be consumed at the cathode. Twice equation 10 is required:

$$Cu(s) = Cu^{2+}(aq) + 2e$$

$$2Ag^{+}(aq) + 2e = 2Ag(s)$$

$$2Ag^{+}(aq) + Cu(s) = 2Ag(s) + Cu^{2+}(aq) \quad (8)$$

This simple result should not be allowed to obscure a very important point. If it is brought about in an electrochemical cell, the rather unremarkable reaction in equation 8 can be used to generate electricity – and electricity can be used to '*do work*'. (This fact is exploited every time a battery is used to turn the starter-motor in a car.) On the other hand, this capacity to 'do work' certainly does not manifest itself when a piece of copper is dropped into silver nitrate solution: yet the same overall reaction takes place.

This again points to a significant difference between the two ways of doing a reaction – the direct (thermal) path and the indirect (electrochemical) one. But what is the underlying reason for this difference? Or, to pose the question in a rather different way:

Question 1 How much work can be extracted from a particular reaction, and how does this depend on the way in which that reaction is carried out?

This is the first of the questions referred to at the beginning of this Section. For now, we put it to one side while we go on to examine another important property of electrochemical cells.

SAQ 2 Look again at Figure 4. Can you now see why it is important to separate the cell components into two half-cells?

2.2 The cell potential

If a self-driving electrochemical cell behaves like a battery, then there must be a 'voltage', that is, a **difference in electrical potential** – or **potential difference** – between the two electrodes. For the moment, we shall refer to it as the cell potential, and note that it is measured in the SI unit volt, V (where $1\ V = 1\ J\ A^{-1}\ s^{-1} = 1\ J\ C^{-1}$). The thermodynamics of electrochemical reactions revolves around the measurement, and interpretation, of such cell potentials. But there is an important proviso.

The point is amply demonstrated by the example in Section 2.1. Concentrate on Figure 5. When current is allowed to flow in the external circuit of the cell, the underlying cell reaction proceeds inside it: electrons dumped at the copper electrode by the oxidation of copper atoms are thereby delivered to the silver electrode, where they reduce Ag^+(aq) ions. Clearly, this *changes* the concentrations of the cell components – that of Cu^{2+}(aq) increasing, while that of Ag^+(aq) drops. But the description of a *changing* system like this is beyond the realm of thermodynamics: rather, the potential difference must be measured when the spontaneous reaction is *prevented* from taking place so that the cell is 'held' at a particular and *constant* composition. In other words, the cell potential must be measured under conditions that permit *no* current to flow between the electrodes. Then, and only then, is the potential a thermodynamic quantity, called the **electromotive force** or **emf**, and given the symbol E.

2.2.1 Measuring the cell emf

There are two methods for measuring the emf. The modern one amounts to replacing the ammeter in Figure 4 with an electronic digital voltmeter – or **digital voltmeter**, for short – which draws a current of only 10^{-6} A, or thereabouts. The older technique, indicated in Figure 6, is based on a very simple idea: it involves including in the external circuit another, and variable, source of potential that can be *balanced against* the potential of the cell being studied – the test cell. In practice, this is achieved by means of a battery and an instrument called a potentiometer: we shall not dwell on the experimental details. Put simply, the basic procedure is to adjust the external source to give a zero current reading on a sensitive ammeter included in the circuit (Figure 6a). At this balance (or *null*) point, the potential of the external source must be equal to that of the test cell so that the two potentials effectively 'cancel' one another: no current flows, and the spontaneous cell reaction is prevented from taking place.

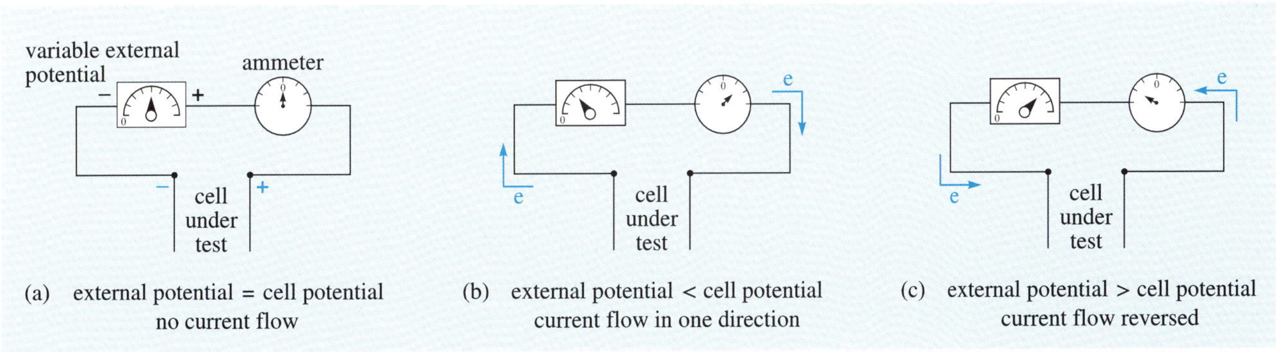

Figure 6 Using a potentiometer to measure the emf of an electrochemical cell: (a) balanced, no current flow; (b) cell 'discharging'; (c) cell 'charging'.

- What is the essential requirement for a null point to be found?
- The cell and the external source must be connected in *opposition* to one another. In other words, the negative terminal of the source must be connected to the negative electrode (anode) of the cell under test, as indicated in Figure 6a. Otherwise it would be impossible to balance the potentials.

It follows that this technique serves to determine *both* the cell potential *and* its *polarity* – that is, which electrode is the anode and which the cathode.

- Why is it important to know the polarity of an electrochemical cell?
- A knowledge of the cell polarity identifies the spontaneous process at each electrode – whether it is an oxidation or a reduction, that is. Hence, it also serves to identify the *direction* of the spontaneous cell reaction – more on which shortly.

When measuring a cell potential, it is obviously important to check that the balance point has been located accurately. With some cells this is easily done using the second technique outlined above. A slight imbalance of the potentials leads to an appreciable flow of current in one direction (Figure 6b) or the other (Figure 6c). Many cells, however, produce very little current in the vicinity of the balance point, so this change in the direction of current flow is hard to locate, and an accurate emf value cannot be determined directly. The problem stems from the *kinetics* of the electrode reactions and is taken up in Section 7.

2.3 Cell diagrams and the sign convention for the cell emf

Consider again the example discussed in Section 2.1: the cell is repeated in Figure 7, only this time we have specified the concentrations of the two electrolytes (0.1 mol dm^{-3}).

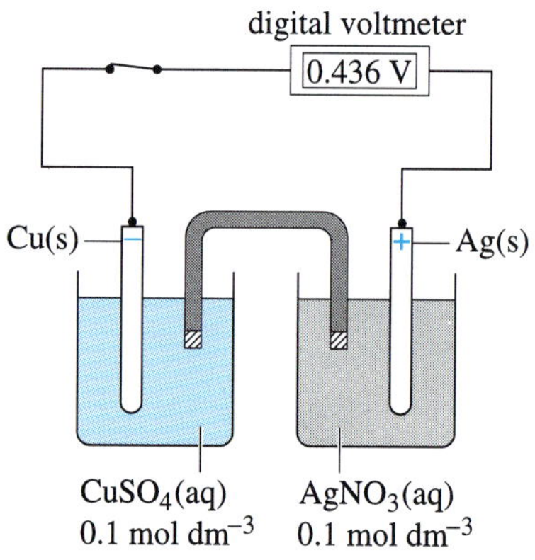

Figure 7 Measuring the cell emf with a digital voltmeter.

Under the (zero current) conditions outlined above, the potential difference between the electrodes of this cell is found to be 0.436 V, copper being the anode and silver the cathode. We should again stress that, under these conditions, no *net* reaction actually occurs. Nevertheless, the observation that copper is the anode indicates that the underlying cell reaction (equation 8) has a spontaneous *tendency* to proceed from left to right, as written:

$$Cu(s) + 2Ag^+(aq) = Cu^{2+}(aq) + 2Ag(s) \tag{8}$$

Clearly, the observed polarity of a cell is an important piece of information. It is recorded in the literature by attaching a *sign* to the corresponding cell emf. Before describing the accepted sign convention, however, it is useful to introduce a more succinct notation for electrochemical cells.

2.3.1 Cell diagrams

A picture such as Figure 7 depicts an electrochemical cell in a rather cumbersome way: it is a *representation* of the cell. The essential features of this arrangement can be conveyed in a simpler, and more convenient way, by means of the following **cell diagram**:

$$Cu(s)|Cu^{2+}(aq)|Ag^{+}(aq)|Ag(s) \qquad (11)$$

Box 2 Cell diagrams and the implied cell reaction

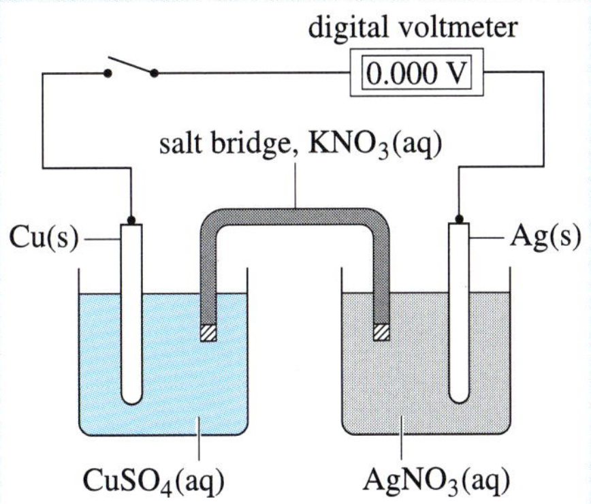

	Left-hand half-cell	*Right-hand half-cell*
Active components:	$Cu(s)$ and $Cu^{2+}(aq)$	$Ag(s)$ and $Ag^{+}(aq)$
Represent as:	$Cu(s)\|Cu^{2+}(aq)$	$Ag^{+}(aq)\|Ag(s)$
	reduced state | oxidized state	oxidized state | reduced state

combine half-cells

↓

Cell diagram: $Cu(s)|Cu^{2+}(aq)|Ag^{+}(aq)|Ag(s)$

Implied cell reaction: $Cu(s) + 2Ag^{+}(aq) = Cu^{2+}(aq) + 2Ag(s)$

The **implied cell reaction** is always written *as if* the process at the left-hand electrode (LHE) in the cell diagram is an oxidation, whereas that at the right-hand electrode (RHE) is a reduction. (Remember **R** for **R**ight and **R**eduction.)

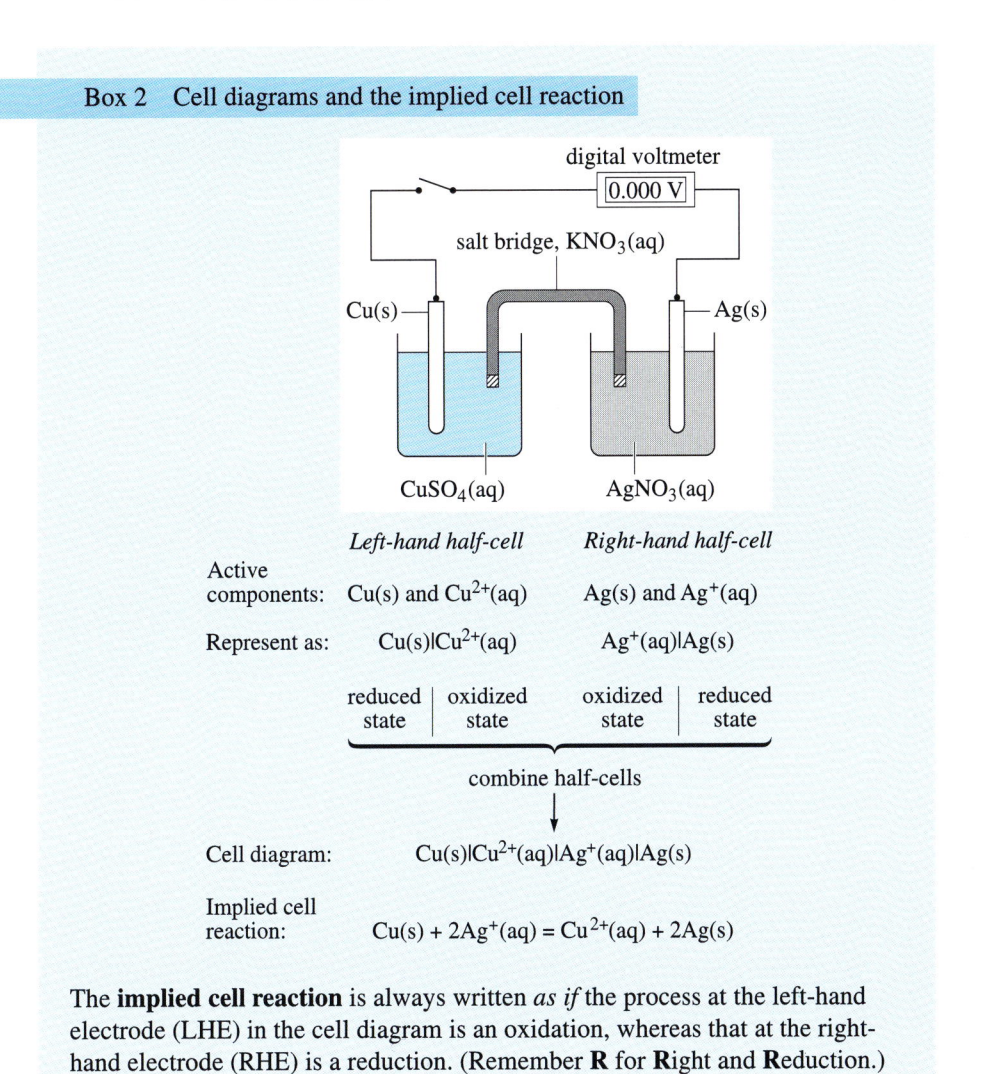

The steps whereby we arrived at this particular cell diagram are included in Box 2. There are several general points to note.

1 A cell diagram effectively represents a juxtaposition of the *active* components of the two half-cells that, together, make up the cell. Ingredients that do not participate in the cell reaction (the sulfate and nitrate ions in this instance) are *not* included – or not usually: there are a few exceptions to this general rule (see Section 3 for an example).

2 The cell diagram takes no account of the actual *stoichiometry* of the cell reaction (cf. cell diagram 11 and equation 8).

3 In each half-cell, the vertical bar represents a *phase boundary* (essentially a physical boundary) between one medium and another. Thus, the left-hand bar in cell diagram 11 represents the boundary between solid copper metal and the aqueous solution of copper ions: the right-hand bar has a similar role. The meaning of the central bar is a little more tricky. With the arrangement in Figure 7, it could be argued

that there are actually *two* phase boundaries between the two solutions; one at either end of the salt bridge. Strictly speaking, the presence of a salt bridge (containing saturated potassium nitrate solution, say) should be indicated in the cell diagram, as:

$$Cu(s)|Cu^{2+}(aq)|\begin{matrix}\text{bridging solution}\\ \text{of } KNO_3(\text{sat.})\end{matrix}|Ag^{+}(aq)|Ag(s)$$

or more simply:

$$Cu(s)|Cu^{2+}(aq)||Ag^{+}(aq)|Ag(s)$$

where the double bar || represents the salt bridge.

Despite this reservation, we shall continue to use the simpler notation throughout this Block (and the next). Thus, you should insert a *single* vertical bar whenever there is a boundary between two components of the cell that are active in the cell reaction.

4 The active ingredients in each half-cell should be written in a particular order. As indicated in Box 2, in the left-hand half-cell, the reduced state (Cu, s) comes first – followed by the oxidized state (Cu^{2+}, aq). The order is *reversed* in the right-hand half-cell.

5 Given the convention in point 4, notice that the cell diagram can be 'read' (from left to right) as *implying* that the process at the left-hand electrode (LHE) is an oxidation (reduced state → oxidized state), whereas that at the right-hand electrode (RHE) is a reduction (oxidized state → reduced state). This way of interpreting the cell diagram is given formal expression via the **implied cell reaction**: the rule for writing this down is included in Box 2. In practice, there's a very simple way to get this right. Given a cell diagram, always write down the *first* species that appears in the diagram (that is, on the extreme left) as the first species in the implied cell reaction: the rest follows.

In applying the rules outlined above, the starting point was the arrangement in Figure 7 – with the copper half-cell on the left, and the silver half-cell on the right. But just suppose that we had chosen to depict the cell the other way about – by way of the following cell diagram:

$$Ag(s)|Ag^{+}(aq)|Cu^{2+}(aq)|Cu(s)$$

- What is the implied cell reaction in this case?

- According to the rule in Box 2, the implied cell reaction must now be the *reverse* of equation 8,

$$2Ag(s) + Cu^{2+}(aq) = 2Ag^{+}(aq) + Cu(s) \quad (12)$$

Now, the experimental behaviour of a real cell must obviously be independent of its orientation on the bench, and of how we choose to depict it on paper. The important general point is that a cell diagram, *in itself*, says nothing about the *direction* of the implied cell reaction. Whether that reaction has a spontaneous tendency to proceed from left to right, or vice versa, depends on the *observed* polarity of the cell. This is why we emphasized the words *as if* in Box 2.

2.3.2 The sign convention for the cell emf

By convention, the emf of a cell – *as represented by a particular cell diagram* – is taken to be the difference:

$$E = \phi(\text{RHE}) - \phi(\text{LHE}) \quad (13)$$

where ϕ (Greek 'phi') is the symbol for electrical potential.* According to electrostatics, the electrical potential at some point is *defined* as the work done in bringing a unit positive charge from infinity to that point. We shall not explore the

* Unfortunately, the symbol ϕ is also used to represent the 'work function', as discussed in Block 6: the distinction between the two uses should be clear from the context.

ramifications of this definition here. Rather we simply note that it carries an important implication in the present context: the electrical potential of a positively charged electrode is *higher* than that of a negatively charged electrode (Figure 8). Thus, *for a 'self-driving' electrochemical cell,*

$$\phi(\text{cathode}) > \phi(\text{anode}) \quad (14)$$

This line of reasoning underlies the convention that is used to record the observed polarity of a cell – by attaching a sign to the measured emf: the convention is spelt out in Box 3.

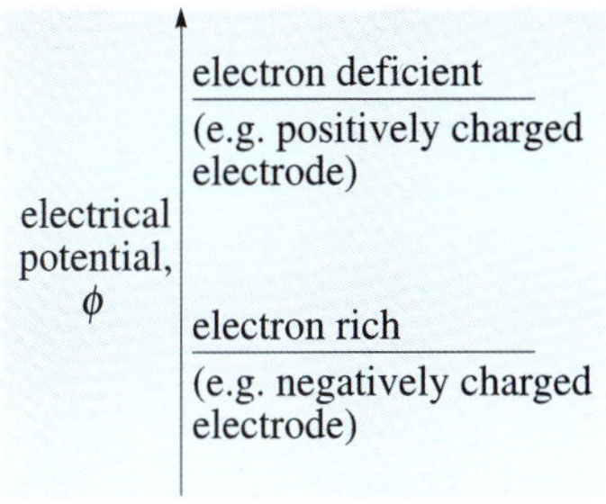

Figure 8 Put simply, like charges repel; unlike charges attract. When a positive test charge is brought up to the positively charged electrode of a cell, more work has to be done than when it is brought up to the negatively charged electrode, and so we report that the electrical potential of the former is higher than that of the latter.

Box 3 The sign convention for the cell emf

For a cell as represented by a particular cell diagram:

$E > 0$ The emf is reported as a positive quantity *if the left-hand electrode in the cell diagram is found to be the anode*, such that $\phi(\text{RHE}) > \phi(\text{LHE})$. Then the implied cell reaction has a tendency to go left $\rightarrow$ right, and electrons would flow left $\rightarrow$ right in the external circuit (that is, both in the direction indicated by >).

$E < 0$ The emf is reported as a negative quantity *if the right-hand electrode in the cell diagram is found to be the anode*, such that $\phi(\text{RHE}) < \phi(\text{LHE})$. Then the implied cell reaction has a tendency to go from right to left, better written as left $\leftarrow$ right, and electrons would flow left $\leftarrow$ right in the external circuit (that is, both in the direction indicated by <).

The crucial point to appreciate is that the sign of an emf is meaningless *unless it is related to a particular cell diagram as written (or drawn) on the page.*

- For the conditions specified in Figure 7, what is the emf of the cell as represented by the following cell diagram?

  $Ag(s)|Ag^+(aq)|Cu^{2+}(aq)|Cu(s)$

- For the conditions in Figure 7, copper is the anode. This is the *right-hand* electrode in the cell diagram, so $\phi(\text{RHE}) < \phi(\text{LHE})$ and E is negative, that is $E = -0.436$ V.

- What is the *sign* of the right-hand electrode in this case?

- It is the negative electrode – the anode.

This suggests an alternative interpretation of the sign convention in Box 3:

The sign of the emf of a cell can be defined as the sign of the electrode on the right in the cell diagram.

STUDY COMMENT The conventions, etc. in Boxes 2 and 3 are among the most important in equilibrium electrochemistry. Make sure you have grasped their full significance by working through the following SAQ.

SAQ 3 Consider the electrochemical cell shown in Figure 9.

(a) Draw a cell diagram to represent the arrangement in Figure 9, and write down the reaction implied by your diagram, as a sum of the separate half-reactions.

(b) With ionic concentrations of 1.0 mol dm^{-3}, the tin electrode was found to be the anode, and the cell emf had a magnitude of 0.010 V. What is the sign of the emf associated with your cell diagram? What is the spontaneous cell reaction?

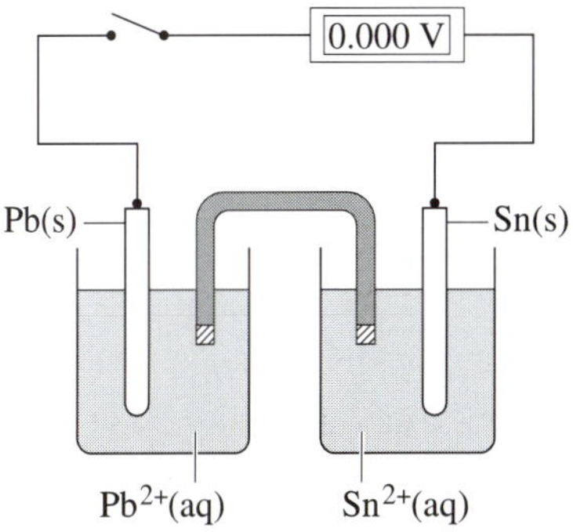

Figure 9 Experimental arrangement for the cell in SAQ 3.

2.4 An electrochemical cell at equilibrium

As noted earlier, cell emfs must be measured under balanced zero-current conditions, because otherwise the underlying reaction would actually take place, leading to changes in the internal composition of the cell. But what effect do such changes have on the measured emf of a cell?

To examine this question, consider again the cell described in SAQ 3, but now represented by the following cell diagram:

$$Sn(s)|Sn^{2+}(aq)|Pb^{2+}(aq)|Pb(s)$$

With ionic concentrations of 1.0 mol dm^{-3}, the measured emf is +0.010 V, and the spontaneous cell reaction is as follows:

$$Sn(s) + Pb^{2+}(aq) = Sn^{2+}(aq) + Pb(s) \quad (15)$$

If a potential is applied across the two electrodes so that it is equal and opposite to the emf of the cell (cf. Figure 6a, Section 2.2.1), then the overall cell reaction cannot take place: the cell is said to be in a state of **imposed equilibrium**.

> Look back at Figure 6b. What do you think happens if the external potential is now reduced slightly, to 0.008 V say?

Electrons begin to flow from left to right in the external circuit: tin goes into solution as the $Sn^{2+}(aq)$ ion, and lead cations, $Pb^{2+}(aq)$, are deposited as metallic lead. Thus, the concentration of $Sn^{2+}(aq)$ ions begins to rise above, and that of $Pb^{2+}(aq)$ ions to fall below, the initial value of 1.0 mol dm^{-3}.

Experiment shows that these changes have the effect of *lowering* the potential of the cell: indeed, as the concentrations continue to change, the cell potential falls from the initial value of 0.010 V and eventually reaches the new applied potential of 0.008 V. At this stage, the *overall* reaction again ceases and a new state of imposed equilibrium is reached. Analysis of the cell gives the concentrations of the two ions corresponding to an emf of 0.008 V. The process of reducing the applied potential could then be repeated to give results like those collected in Table 1. At this stage, we shall not embark on a detailed analysis of these results. Rather we simply note that they raise two further questions.

Table 1 Effect of ion concentration[a] on the emf of the cell: $Sn(s)|Sn^{2+}(aq)|Pb^{2+}(aq)|Pb(s)$.

E/V	$c(Sn^{2+})$/mol dm^{-3}	$c(Pb^{2+})$/mol dm^{-3}
0.010	1.00	1.00
0.008	1.08	0.92
0.006	1.16	0.84
0.004	1.23	0.77
0.002	1.30	0.70
0	1.37	0.63

[a] Throughout this Block we use the symbol c, rather than square brackets [ ], to represent the molar concentration.

First, it is clear that the emf of a cell does indeed depend on the concentrations of its active components, but what is the nature of this concentration-dependence? Or, to put it another way:

> **Question 2** Precisely how does the emf of an electrochemical cell depend on the concentrations of its active components?

In fact, this question is only part of a much broader issue. According to the results in Table 1 (and as demonstrated in the video sequence), normal operation lowers the emf of a cell – eventually to *zero*. Now, with no potential difference between the electrodes, no current will flow, *even if the two electrodes are joined together directly*.

- What, then, has happened to the underlying cell reaction (equation 15)?
- Like any other spontaneous reaction, it must have reached equilibrium.

This is a very important conclusion. When the emf of a cell reaches zero, no *net* reaction occurs: the cell is then in a *true state of equilibrium* with respect to the cell reaction. In other words, the condition $E = 0$ must represent a chemical reaction at equilibrium (albeit in an electrochemical cell), and this is turn suggests strongly that the emf is a *thermodynamic* quantity. So our third question becomes:

Question 3 What is the significance, in thermodynamic terms, of a cell emf different from zero?

2.5 Summary of Section 2

1 The overall reaction in a 'self-driving' electrochemical cell can be written as the sum of the two balanced half-reactions, one an oxidation and the other a reduction, that occur at the anode (negative electrode) and cathode (positive electrode), respectively, of the cell.

2 If the two electrodes are joined together directly, then the spontaneous cell reaction manifests itself as a flow of electrons from the anode to the cathode, in the external circuit.

3 Alternatively, the cell emf E and the cell polarity can be determined under balanced zero-current conditions. The spontaneous direction of the cell reaction can be predicted from the cell polarity.

4 A cell can be represented on paper by a cell diagram. The conventions for allocating a sign to the emf associated with it, and for writing down the implied cell reaction, are summarized in Box 4.

Box 4

1 Draw a cell diagram (see the example in Box 2).

2 Write down the *implied* cell reaction, with oxidation at the LHE, reduction at the RHE.

3 Then:

if the LHE *is* the anode, $E > 0$ and the implied cell reaction has a tendency to go $L \rightarrow R$ as written;

if the LHE *is not* the anode, $E < 0$ and the implied cell reaction has a tendency to go $L \leftarrow R$ as written.

5 Normal operation lowers the emf of a cell: it drops to zero when the underlying cell reaction reaches equilibrium.

In addition, this Section has raised three important questions:

Question 1 How much work can be extracted from a particular reaction, and how does this depend on the way in which that reaction is carried out?

Question 2 Precisely how does the emf of an electrochemical cell depend on the concentrations of its active components?

Question 3 What is the significance, in thermodynamic terms, of a cell emf different from zero?

For the next few Sections, our discussions will centre around these questions – starting with the third one.

SAQ 4 The electrochemical cell represented by the cell diagram below is called a Daniell cell, after its discoverer (in 1836). A cell like this was one of the earliest batteries, and found wide application in telegraph and railway signalling.

$Zn(s)|Zn^{2+}(aq)|Cu^{2+}(aq)|Cu(s)$

(a) Write down the implied cell reaction.

(b) With ionic concentrations of 1.0 mol dm^{-3}, the zinc electrode is found to be the anode. Allocate a sign to the cell emf and indicate the *spontaneous* cell reaction.

(c) Use information from the S342 *Data Book* to calculate $\Delta G_m^\ominus$ at 298.15 K for the implied cell reaction. Does your answer accord with the observed polarity of the cell?

3 EMF AND 'REACTION TENDENCY': A HYPOTHESIS

A clue to the answer to Question 3 lies in the language used throughout Sections 2.3 and 2.4. We spoke of there being a 'tendency' for the spontaneous cell reaction to occur: at equilibrium ($E = 0$), this tendency is no longer apparent – not on a macroscopic scale, at least. Language like this should have a familiar ring! From a thermodynamic point of view, it suggests strongly that the cell emf is an alternative measure of '*reaction tendency*': alternative, that is, to the more familiar arbiter of spontaneous chemical change – the Gibbs function, G. The answer to SAQ 4 contains a hint that this may indeed be so.

To examine the implications of this hypothesis, we turn now to the cell illustrated schematically in Figure 10 (the second of the two cells featured in video band 8).

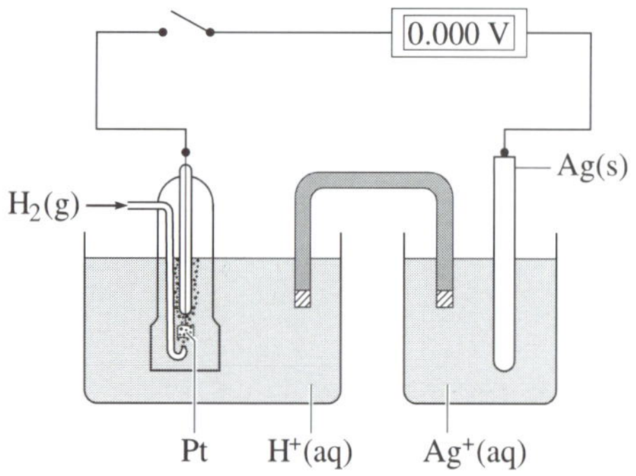

Figure 10 Experimental arrangement for the cell: Pt, $H_2(g)|H^+(aq)|Ag^+(aq)|Ag(s)$.

The arrangement is a little different from the simple combinations of metal/aqueous metal cation half-cells that you met in Section 2. This time, the half-cell on the left-hand side involves a gas – hydrogen – and is called a **hydrogen electrode**: you will see later on that it has a central role to play in electrochemistry. Basically, it comprises a piece of platinum dipping into an aqueous solution containing $H^+(aq)$ ions: hydrogen gas is bubbled steadily over the surface of the platinum. The metal serves both to provide a large surface area, so that the hydrogen gas can equilibrate with the hydrogen ions in the solution, and to allow electrical contact with the external circuit. Although the platinum *does not* take part in the cell reaction, its presence is usually indicated explicitly in the cell diagram, which can be written as follows:

Pt, $H_2(g)|H^+(aq)|Ag^+(aq)|Ag(s)$

- Does the *order* of the active species in the left-hand half-cell accord with the rules in Section 2.3.1?
- Yes. Here, $H_2(g)$ and $H^+(aq)$ are the reduced and oxidized states, respectively.

- Write down the implied cell reaction.
- One way to balance the electrons transferred is as follows:

LHE (oxidation): $\frac{1}{2}H_2(g) = H^+(aq) + e$

RHE (reduction): $Ag^+(aq) + e = Ag(s)$

$$\tfrac{1}{2}H_2(g) + Ag^+(aq) = H^+(aq) + Ag(s) \qquad (16)$$

- Use information from the S342 *Data Book* to calculate $\Delta G_m^\ominus$ at 298.15 K for the reaction in equation 16. What do you conclude from your answer?
- Hopefully, you recognized that equation 16 is just the *reverse* of the formation reaction for $Ag^+(aq)$. (If not, refer back to Block 1 and/or to the notes in Section 2 of your S342 *Data Book*.) Hence,

$\Delta G_m^\ominus(16) = -\Delta G_f^\ominus(Ag^+, aq) = -77.1\ \mathrm{kJ\ mol^{-1}}$

Up till now, we have often used the simple criterion $\Delta G_m^\ominus < 0$ to identify a thermodynamically favourable reaction (at 298.15 K, at least). On this basis (as in the answer to SAQ 4c), we would predict a spontaneous tendency for the reaction to go from left to right, as written. How, then, does this prediction compare with the experimental behaviour of the cell?

Concentrate now on the experimental observations collected in Table 2. Here, the ionic concentrations in the two half-cells have been varied *independently* of one another: in each case the resulting cell polarity (under balanced zero-current conditions) was noted down by attaching a sign to the measured emf.

Table 2 Experimental observations on the cell: Pt, $H_2(g)|H^+(aq)|Ag^+(aq)|Ag(s)$.

$c(H^+)/\mathrm{mol\ dm^{-3}}$	$c(Ag^+)/\mathrm{mol\ dm^{-3}}$	Cell emf
1.0	1.0	$E > 0$
0.1	0.1	$E > 0$
1.0×10^{-5}	0.1	$E > 0$
1.0×10^{-5}	1.0×10^{-21}	$E < 0$

To what extent do these observations tie in with the prediction above? Think carefully about this question before reading on.

Concentrate on equation 16. According to the sign convention introduced in Section 2.3.2 (Box 3), a positive cell emf does indeed indicate a tendency for this reaction to go from left to right, as written. Thus, it is tempting simply to equate the two conditions $E > 0$ and $\Delta G_m^\ominus < 0$ as alternative criteria for a spontaneous chemical reaction.

But there is a problem. As the final entry in Table 2 shows, a drastic reduction in the concentration of $Ag^+(aq)$ actually *reverses* the polarity of the cell:* the cell emf becomes negative, and the reaction then has a tendency to proceed in the opposite direction (that is, from right to left, as written). At first sight, the thermodynamic criterion we have used so far contains no hint of this possibility: indeed, it *appears* to

* As shown in video band 8, this can be achieved by adding sulfide ions, $S^{2-}(aq)$, and hence precipitating out the sparingly soluble silver salt, Ag_2S. Of course, such small concentrations can't actually be measured, but they can be *calculated* – from the solubility product of the sparingly soluble salt in question. This sort of calculation is taken up in Section 4.7.

predict that the reverse reaction cannot happen of its own accord, because $\Delta G_m^\ominus$ must then be positive (that is, +77.1 kJ mol^{-1}).

In fact, the problem is more apparent than real: it stems from an overly simplistic interpretation of the sign of $\Delta G_m^\ominus$ for a reaction. As we reminded you in Block 1, $\Delta G_m^\ominus$ is really just a measure of the standard equilibrium constant $K^\ominus$ for a reaction (at a particular temperature of course):

$$\Delta G_m^\ominus = -RT \ln K^\ominus \quad (17)$$

- According to this relation, what is the actual implication of a negative value for $\Delta G_m^\ominus$?

- It simply implies that the equilibrium constant must be greater than one.

The following SAQ contains a clue to how this more critical interpretation of $\Delta G_m^\ominus$ may resolve our dilemma.

SAQ 5 Consider a simple isomerization reaction of the type

$$A(aq) = B(aq)$$

If $\Delta G_m^\ominus = -5$ kJ mol^{-1} for this reaction at 298.15 K, what is the corresponding value of the equilibrium constant?

Suppose now that A and B are mixed in proportions such that $c_B = 100c_A$, where c represents the concentration. What spontaneous change would you expect to occur?

The example in SAQ 5 suggests that our measure of 'reaction tendency' is in some sense incomplete: evidently, it should take account *both* of the value of $\Delta G_m^\ominus$ (and hence the size of the equilibrium constant) and of the actual composition of the system. The importance of this second factor certainly ties in with the observed behaviour of the cell: after all, the spontaneous cell reaction was reversed by manipulating the concentrations of the cell components. The problem that remains, then, is to find a precise, quantitative way of taking both factors into account.

To this end, we embark now on a closer examination of the thermodynamics of chemical change *in general*. Inevitably, this represents a fairly substantial hiatus in your study of electrochemistry. Bear with us. Remember that our aim is two-fold: first, to find a true thermodynamic measure of 'reaction tendency', and second to link this with the electrochemical criterion, $E > 0$, established above.

4 THE THERMODYNAMICS OF CHEMICAL CHANGE

4.1 Introduction

The problem identified in the previous Section is not restricted to reactions in electrochemical cells. Rather, it is part of the broader problem of predicting the direction of spontaneous change in *any* chemical system. As you saw in the Second Level Inorganic Course, the solution lies in a careful application of the **second law of thermodynamics** to a typical chemical system: a reaction taking place under the normal laboratory conditions of constant temperature and pressure. This analysis resulted in a simple criterion for a spontaneous reaction; it was written in terms of the Gibbs function as follows:

$$\Delta G < 0 \quad \text{(at constant } T \text{ and } p\text{)} \tag{18}$$

But what do we mean by 'ΔG' in this context? In the Second Level Inorganic Course, we went on to assert that $\Delta G_m^{\ominus}$ is the molar value of ΔG when reactants and products are in their 'standard' states. This, in turn, suggests an alternative interpretation of our problem: perhaps the criterion $\Delta G_m^{\ominus} < 0$ is incomplete because it is restricted to reactions 'under standard conditions'. But what are these standard conditions, and how can we lift this restriction in order to find a more general measure of 'reaction tendency'?

To tackle these questions, our first step will be to formulate inequality 18 in a more precise way. This, together with the definition of a further thermodynamic quantity – the chemical potential – will eventually provide us with a recipe for determining the 'reaction tendency' in any chemical system, be it an electrochemical cell or a simple gaseous reaction, no matter what its composition.

STUDY COMMENT We are concerned here (and in Sections 4.2 to 4.6 in particular) with developing a number of basic thermodynamic relations: don't worry if you sometimes find the argument a bit hard-going. You will not be expected to reproduce the derivations, only to be able to use the final results. For this reason, it is particularly important that you attempt the SAQs as you go along – and Exercise 2 when you have completed the Section. Start by working through Exercise 1: it revises and draws together a number of important ideas that were introduced in Blocks 1 and 2, and provides the essential background to our discussion here.

EXERCISE 1 (revision)

Part 1 (*based on Block 1*) Consider the following gaseous reaction:

$$N_2O_4(g) = 2NO_2(g) \tag{19}$$

where, at 298.15 K, $\Delta H_m^{\ominus} = 57.2\ \text{kJ mol}^{-1}$ and $\Delta S_m^{\ominus} = 175.9\ \text{J K}^{-1}\ \text{mol}^{-1}$.

(a) Use this information to calculate $\Delta G_m^{\ominus}$ at 325 K, and the corresponding values of $K^{\ominus}$ and K_p, for the reaction in equation 19. *State any assumptions involved in your calculations.*

(b) Derive an expression for K_p in terms of the equilibrium yield y of NO_2, where $y = p(NO_2)/p_{tot}$. Use this expression to confirm that $y = 0.616$ at 325 K, when the overall pressure $p_{tot} = p^{\ominus} = 1$ bar. In what way does the value of y describe the *composition* of the equilibrium mixture?

Part 2 (*based on Block 2*) A general, and very convenient, measure of how far a reaction has gone is the **extent of reaction**, ξ. This quantity was introduced in a kinetic context in Block 2, and was defined as follows:

$$\xi = \frac{n_Y - n_{Y,0}}{\nu_Y}$$

where n_Y is the amount of substance Y present when the reaction has proceeded to an extent ξ, $n_{Y,0}$ is the initial amount of that substance, and ν_Y is its stoichiometric number in the balanced reaction equation.

(a) What are the stoichiometric numbers of N_2O_4 and NO_2 in equation 19?

(b) If we start with one mole of N_2O_4, and this is *completely* converted into NO_2, what are the initial and final values of ξ for the reaction in equation 19?

(c) Suppose now that ξ_e is the *extent of reaction at equilibrium* for the conditions specified in Part 1, namely $T = 325$ K and $p_{tot} = 1$ bar. Assuming 1 mol of N_2O_4 is initially present, use the definition of ξ to write expressions for $n(N_2O_4)$, $n(NO_2)$ and $n_{tot} = n(N_2O_4) + n(NO_2)$ *at equilibrium* in terms of ξ_e. Hence determine the value of ξ_e corresponding to the equilbrium yield $y = 0.616$.

4.2 The Gibbs 'valley'

The basic implication of inequality 18 is very simple: a reaction will happen if, and only if, there is a decrease in the Gibbs function of the system. This idea was given visual expression in the Second Level Inorganic Course: spontaneous reactions were seen to 'roll downhill' on a landscape where 'altitude' is measured by the Gibbs function (Figure 11).

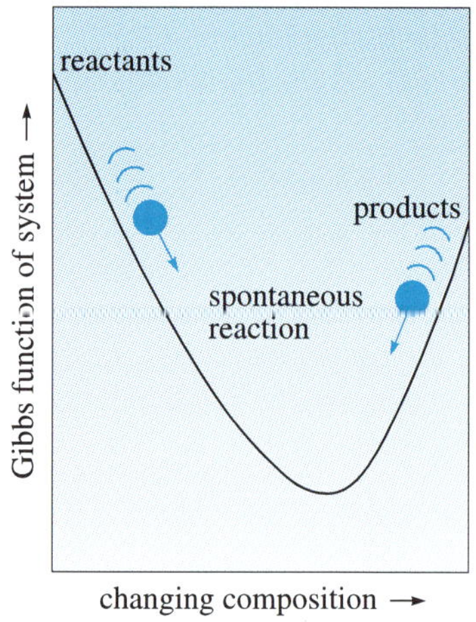

Figure 11 The direction of spontaneous change is down to the valley of the Gibbs function.

To make this idea more precise, consider again the decomposition reaction in equation 19:

$$N_2O_4(g) = 2NO_2(g) \qquad (19)$$

For a simple gaseous reaction like this, results we shall introduce later on can be used to *calculate* the Gibbs function of the system (under specified conditions) as a function of the extent of reaction ξ. Figure 12 shows the variation in G for the conditions examined in Exercise 1 (that is, a constant temperature of 325 K and a constant pressure of 1 bar). Notice that the extent of reaction runs from zero (no reaction – assumed to be 1 mol of pure N_2O_4 in this case) to one (complete dissociation into NO_2).

To begin with, concentrate on the left-hand side of the curve in Figure 12. Here, the starting point (that is, $\xi = 0$) represents pure N_2O_4. As ξ increases, the Gibbs function of the system decreases, until it reaches a minimum value – at the point labelled (b) in Figure 12.

- What is the significance of the minimum in the curve?

- As you should have found in answering Exercise 1, at this point $\xi = \xi_e = 0.445$ mol: it corresponds to the position of *equilibrium* for the reaction in equation 19 (under the conditions implied by Figure 12).

Put another way, under these conditions there is a tendency for N_2O_4 to dissociate spontaneously into NO_2: the composition of the system changes until it reaches equilibrium – the bottom of the Gibbs 'valley'. Our reformulation of inequality 18 follows from a more precise specification of the conditions at this point.

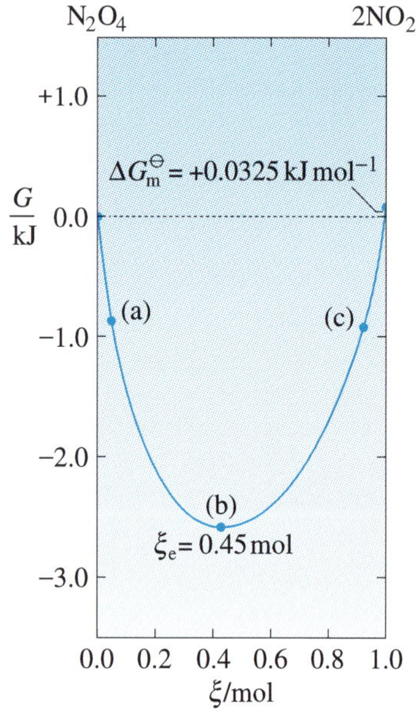

Figure 12 The variation in G with extent of reaction ξ for the system $N_2O_4(g) = 2NO_2(g)$, at a constant temperature of 325 K and a constant pressure of 1 bar. (Under these conditions, $\Delta G_m^\ominus \approx 0$ kJ mol^{-1}, $K^\ominus \approx 1$, and so the equilbrium position is *roughly* half-way. We chose these conditions to make this plot as clear as possible.)

- Draw in the tangent to the curve at point (b) in Figure 12. What do you conclude about the slope of the curve at this point?

- Your tangent should be *parallel* to the ξ-axis: the slope of the curve must, therefore, be zero at this point.

  Try to express this result in symbols. Refer back to Section 3 in the AV Booklet if you are uncertain how to do this.

The slope of the plot of one variable against another can be written as a *differential*. In this case, the variables are G and ξ, and the slope at the minimum (that is, at equilibrium) is zero, so the required expression is as follows:

$$\frac{dG}{d\xi} \text{ or } (dG/d\xi) = 0 \text{ at equilibrium} \qquad (20)$$

Equation 20 is a formal statement of the condition for equilibrium in a chemical reaction, under the normal laboratory conditions of constant temperature and pressure.

Look again at Figure 12. What is the *sign* of $(dG/d\xi)$ on either side of the minimum? Can you now suggest a reformulation of inequality 18?

According to Figure 12, the slope of the curve changes sign at the minimum: it is negative on one side (at point (a), for example) but positive on the other (at point (c), say). Now a reaction proceeds in the direction of decreasing G: it follows that the reaction in equation 19 will proceed from left to right (that is, in the direction of increasing ξ) if, and only if,

$$(dG/d\xi) < 0 \qquad (21)$$

This, then, is our reformulation of the criterion for spontaneous change in a chemical system. Notice that it has an important corollary: if $(dG/d\xi) > 0$, then the *reverse* reaction should take place. In terms of the picture in Figure 12, the system then 'rolls down' the other side of the Gibbs 'valley'.

To summarize: the sign of $(dG/d\xi)$ is the true arbiter of spontaneous change in a chemical system; at equilibrium, $(dG/d\xi) = 0$.

But there remain two questions. Is it possible to determine the sign of $(dG/d\xi)$ for a given chemical system, and how is this related (as the example in SAQ 5 suggests that it must be) to the value of $\Delta G_m^\ominus$ for the reaction in question? One way to tackle these questions is to introduce another thermodynamic quantity – the **chemical potential**.

4.3 The chemical potential

The chemical potential of a *pure* substance is just the *molar* Gibbs function G_m of that substance: this follows from the *definition*

$$G = n\mu \qquad (22)$$

where μ (Greek 'mu') denotes the chemical potential and n is the amount of the substance in question. Thus, when $n = 1$ mol, $G/n = G_m = \mu$.

- According to this definition, what is the SI unit of μ?
- The SI unit of G is J, and that of n is mol, so the SI unit of μ is J mol^{-1} (joules per mole).

The Gibbs function for a *mixture of substances* can be written as a *sum* of such terms, one for each substance present.

Consider again the reaction in equation 19:

$$N_2O_4(g) = 2NO_2(g) \qquad (19)$$

- Write an expression for G for a mixture containing amounts $n(N_2O_4)$ and $n(NO_2)$ of N_2O_4 and NO_2, respectively.
- $$G = \mu(N_2O_4)n(N_2O_4) + \mu(NO_2)n(NO_2) \qquad (23)$$

Now, our aim is to determine the value of $(dG/d\xi)$ at some particular point on a curve like the one in Figure 12 (point (a), say). With this in mind, suppose that equation 23 represents the value of G at such a point. The amounts of N_2O_4 and NO_2 must then be related to the extent of reaction, ξ, at that point. This follows from the formal definition of ξ (see Exercise 1), which can be rearranged to read:

$$n_Y = n_{Y,0} + \nu_Y\xi \qquad (24)$$

Suppose now that the Gibbs function of the system changes as a result of a tiny change in the extent of reaction, from ξ_1 to ξ_2 (say), such that:

$$\Delta\xi = \xi_2 - \xi_1 \qquad (25)$$

Before reading further, try to work out expressions for $\Delta n(N_2O_4)$ and $\Delta n(NO_2)$, the corresponding changes in the amounts of N_2O_4 and NO_2. Concentrate on equation 24.

From the definition in equation 24 (and recalling $\nu(N_2O_4) = -1$),

$$\begin{aligned}\Delta n(N_2O_4) &= n_2(N_2O_4) - n_1(N_2O_4)\\ &= \{n_0(N_2O_4) - \xi_2\} - \{n_0(N_2O_4) - \xi_1\}\\ &= -(\xi_2 - \xi_1)\\ &= -\Delta\xi = \nu(N_2O_4)\Delta\xi \end{aligned} \tag{26}$$

Similarly,

$$\Delta n(NO_2) = +2\Delta\xi = \nu(NO_2)\Delta\xi \tag{27}$$

Now, imagine a situation where ξ changes by an *infinitesimal* amount. In this limit, we replace the finite increment Δ by the infinitesimal increment d (again, refer back to Section 3 in the AV Booklet if you are uncertain about this notation), so that equations 26 and 27 become

$$\mathrm{d}n(N_2O_4) = -\mathrm{d}\xi \tag{28}$$

and

$$\mathrm{d}n(NO_2) = +2\mathrm{d}\xi \tag{29}$$

But what is the corresponding change dG in the Gibbs function of the system? As equation 23 suggests, this is related to the change in the *amounts* of N_2O_4 and NO_2: at constant temperature and pressure, it is given by*

$$\mathrm{d}G = \mu(N_2O_4)\mathrm{d}n(N_2O_4) + \mu(NO_2)\mathrm{d}n(NO_2) \quad (\text{constant } T \text{ and } p) \tag{30}$$

Substituting the expressions in equations 28 and 29 into equation 30 leads to

$$\begin{aligned}\mathrm{d}G &= \mu(N_2O_4)(-\mathrm{d}\xi) + \mu(NO_2)(+2\mathrm{d}\xi)\\ &= \{2\mu(NO_2) - \mu(N_2O_4)\}\mathrm{d}\xi\end{aligned}$$

which can be reorganized into

$$(\mathrm{d}G/\mathrm{d}\xi) = 2\mu(NO_2) - \mu(N_2O_4) \quad (\text{constant } T \text{ and } p) \tag{31}$$

Equation 31 represents an important step towards our objective: it shifts attention from the Gibbs function of the *system* to the chemical potentials of its individual components. Thus, our criterion for spontaneous change can now be couched in different terms (Figure 13), as summarized by statements 32–34.

If $\mu(N_2O_4) > 2\mu(NO_2)$, then $(\mathrm{d}G/\mathrm{d}\xi) < 0$ and the reaction goes from left to right, as written. (32)

If $\mu(N_2O_4) < 2\mu(NO_2)$ then $(\mathrm{d}G/\mathrm{d}\xi) > 0$ and the reaction goes from right to left, as written. (33)

But, when $\mu(N_2O_4) = 2\mu(NO_2)$, then $(\mathrm{d}G/\mathrm{d}\xi) = 0$ and the reaction is at equilibrium. (34)

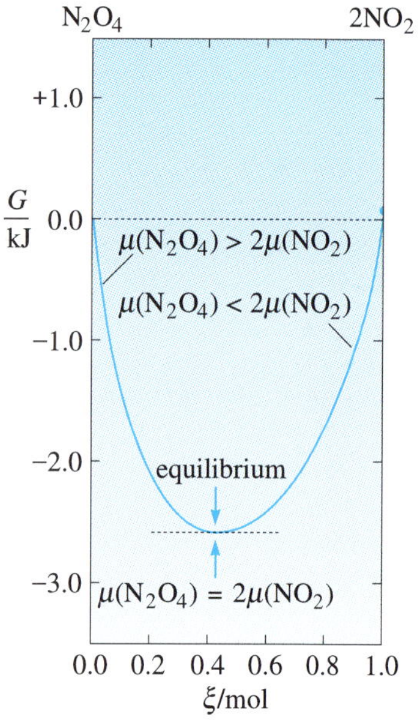

Figure 13 The direction of spontaneous change is determined by the *sign* of the slope of the 'Gibbs landscape'; for reaction 19, $(\mathrm{d}G/\mathrm{d}\xi) = 2\mu(NO_2) - \mu(N_2O_4)$.

Equation 31 is just one example of a completely general result. Notice that the right-hand side of this equation is actually the *sum* of μs for the substances present, *each multiplied by the appropriate stoichiometric number* (from equation 19): *the minus sign derives from the convention that ν is negative for a reactant.*

* If you have studied calculus *beyond* the level assumed in this Course, then you may recognize that differentation of the expression in equation 23 strictly gives:

$$\begin{aligned}\mathrm{d}G = &\{\mu(N_2O_4)\mathrm{d}n(N_2O_4) + \mu(NO_2)\mathrm{d}n(NO_2)\}\\ &+ \{n(N_2O_4)\mathrm{d}\mu(N_2O_4) + n(NO_2)\mathrm{d}\mu(NO_2)\}\end{aligned}$$

However, it is a further consequence of the laws of thermodynamics that the second term in curly brackets is *zero* at constant T and p: equation 30 is therefore correct.

Thus, for a general reaction, which can be written in a more user-friendly 'alphabetical' form as follows:

$$aA + bB + \ldots = pP + qQ + \ldots$$

equation 31 becomes

$$\begin{aligned}(dG/d\xi) &= (\nu_P\mu_P + \nu_Q\mu_Q + \ldots) + (\nu_A\mu_A + \nu_B\mu_B + \ldots)\\ &= (p\mu_P + q\mu_Q + \ldots) - (a\mu_A + b\mu_B + \ldots) \qquad (35)\end{aligned}$$

since $\nu_P = p$, $\nu_Q = q$,... etc., *but* $\nu_A = -a$, $\nu_B = -b$, and so on.

STUDY COMMENT If you would like to practise using equation 35 at this stage, try the following SAQ.

SAQ 6 Use equation 35 to write expressions for $(dG/d\xi)$ for each of the following reactions:

(a) $\frac{1}{2}H_2(g) + Ag^+(aq) = H^+(aq) + Ag(s)$

(b) $Zn(s) + Cu^{2+}(aq) = Zn^{2+}(aq) + Cu(s)$

(c) $AgCl(s) = Ag^+(aq) + Cl^-(aq)$

4.4 Standard states and activities

Equation 35 provides a means of predicting the sign of $(dG/d\xi)$ only if it is possible to measure (or otherwise determine) values of the chemical potential. But it isn't: like the Gibbs function itself, and other thermodynamic quantities, it is only possible to determine *differences* in the chemical potential of a particular substance, not absolute values. This may seem a rather disturbing observation, but it can be circumvented.

To begin with, it is important to appreciate that the chemical potential of a substance depends on the conditions, and in particular on the composition of the reaction mixture in which it finds itself. This is implicit in the statements 32–34 above: evidently $\mu(N_2O_4)$ and $\mu(NO_2)$ change as the reaction proceeds, until the equilibrium condition, 34, is satisfied. Thermodynamics enables us to quantify this composition-dependence. In particular, it provides recipes for relating the value of μ under one set of conditions to that under a specified set of **standard conditions**. *At a constant temperature T*, this relation can be expressed quite generally as follows:

$$\mu - \mu^{\ominus} = RT\ln a \qquad (36)$$

where $\mu^{\ominus}$ is the **standard chemical potential** of the substance (that is, the value of μ under standard conditions) and a is called the **activity**.

- According to equation 36, what is the SI unit of a?
- It has no unit. We can take the logarithm of a quantity only if it is a pure number, so a must be dimensionless. (Refer back to Section 1 in the AV Booklet if you are uncertain about this.) More illuminating, perhaps, is the fact that μ, $\mu^{\ominus}$ and RT all have the same unit, J mol^{-1} in SI.

Unfortunately, the derivation of this vital relation lies beyond the scope of this Course, so we must ask you to take it on trust. But to use equation 36 we obviously need to know how the standard conditions (or the **standard state** as it is often called) are defined – and, indeed, what is meant by 'activity'. It turns out that these two concepts are intimately interrelated. The standard state is quite arbitrary. It is chosen to be the most convenient for the job in hand, and selected such that the activity of the substance in question can be related to measurable quantities. To make this clearer, we ask you to accept the following *definitions of the difference* $(\mu - \mu^{\ominus})$ *in three very important cases* (*all at the constant temperature T*).

Case 1 For a pure liquid or solid,

either $\mu - \mu^{\ominus} = 0$ (37)

or, equivalently, $\mu = \mu^{\ominus}$ (38)

There are two ways to interpret this definition. The first, implied by equation 38, is that the chemical potential of a liquid or solid depends only on the temperature: the standard state is simply the pure substance at that temperature.*

- Can you see what the second way is? Compare equations 36 and 37.
- $RT \ln a = 0$ if $a = 1$, *so the activity of a pure liquid or solid is always unity.*

We shall make use of both interpretations.

Case 2 For a pure gas, or gaseous component in a mixture,

$$\mu - \mu^{\ominus} = RT \ln (p/p^{\ominus}) \quad (39)$$

where p is the pressure of the gas, or its partial pressure in a mixture, and $p^{\ominus}$ is the standard pressure, defined to be 1 bar. In this case, $a = p/p^{\ominus}$, and the chemical potential of the gas has its standard value when the pressure $p = p^{\ominus} =$ 1 bar. Alternatively, the activity of a gas is unity when its pressure is 1 bar.

Case 3 For a solute in solution,

$$\mu - \mu^{\ominus} = RT \ln (c/c^{\ominus}) \quad (40)$$

where c is the molar concentration of the solute, and $c^{\ominus}$ is a standard concentration. Throughout this Block (and the next), we shall adopt the standard $c^{\ominus} = 1 \text{ mol dm}^{-3}$. Then, $a = c/c^{\ominus}$, and the chemical potential of the solute has its standard value when $c = c^{\ominus} = 1 \text{ mol dm}^{-3}$ – i.e. the solute is then at unit activity.†

These three cases have been chosen carefully. Case 1 is quite general: indeed, we shall *assume* that it applies also to the *solvent* in a solution. The validity of the definitions in Cases 2 and 3 is more restricted, however: the expressions in equations 39 and 40 represent the *simplest* type of behaviour found experimentally, for gases and for solutions, respectively. Gases obeying equation 39 are called **ideal gases**, and a solution of components that obey equation 40 is termed an **ideal solution**. It follows that the standard state of an ideal gas is established at $p = 1$ bar, and of a solute in an ideal solution when its concentration $c = 1 \text{ mol dm}^{-3}$.

The notion of an ideal gas was introduced in Block 1, and some of the implications of assuming ideal gas behaviour were touched on at that stage: they are explored further in Section 4.6. Equally, the extent to which any *real* solution – and in particular the electrolyte solutions that are our main concern here – can be assumed to behave ideally is examined in detail in Section 4.7. For the moment, however, we shall confine ourselves to the simple expressions above, and see to what extent they provide an answer to our central problem: the determination of $(dG/d\xi)$.

One final point: we are now in a position to give a **formal definition of the standard molar Gibbs function change for a reaction**, $\Delta G_m^{\ominus}$. It is defined in terms of the

* Strictly speaking, the chemical potential of a liquid or a solid (and of a solute in solution) also depends very slightly on the pressure. The standard state then requires specification of the standard pressure ($p^{\ominus} = 1 \text{ bar} = 10^5$ Pa). For pressures little diferent from ambient, however, the effect is negligible, and we shall not consider it further in this Course.

† If you consult other texts, you may find this definition expressed in terms of an alternative measure of concentration – the *molality*, symbol m. Molality is defined as the amount of solute per *kilogram* of solvent, whence $m^{\ominus} = 1 \text{ mol kg}^{1}$. It is thus independent of temperature (unlike the molar concentration c), and is also more easily determined with accuracy than is c. For the *dilute aqueous* solutions we shall be concerned with here, the two measures are very similar: the difference between them is significant only in the most accurate work.

standard chemical potentials of the reactants and products by the following expression (cf. equation 35):

$$\Delta G_{\text{m}}^{\ominus} = (\nu_{\text{P}}\mu_{\text{P}}^{\ominus} + \nu_{\text{Q}}\mu_{\text{Q}}^{\ominus} + \ldots) + (\nu_{\text{A}}\mu_{\text{A}}^{\ominus} + \nu_{B}\mu_{\text{B}}^{\ominus} + \ldots)$$

$$= (p\mu_{\text{P}}^{\ominus} + q\mu_{\text{Q}}^{\ominus} + \ldots) - (a\mu_{\text{A}}^{\ominus} + b\mu_{\text{B}}^{\ominus} + \ldots) \qquad (41)$$

where, as before, the stoichiometric coefficients of the reactants, ν_{A}, ν_{B}, ..., are negative numbers – that is, $\nu_{\text{A}} = -a$, $\nu_{\text{B}} = -b$, etc.

4.5 The calculation of 'reaction tendency'

To gain some familiarity with the definitions above, consider again the reaction in equation 19:

$$N_2O_4(g) = 2NO_2(g) \qquad (19)$$

For this system:

$$(dG/d\xi) = 2\mu(NO_2) - \mu(N_2O_4) \qquad (31)$$

As a first step, the general expression in equation 36 ($\mu = \mu^{\ominus} + RT\ln a$) can be used to expand the right-hand side of this equation:

$$(dG/d\xi) = 2\{\mu^{\ominus}(NO_2) + RT\ln a(NO_2)\} - \{\mu^{\ominus}(N_2O_4) + RT\ln a(N_2O_4)\}$$

$$= \{2\mu^{\ominus}(NO_2) - \mu^{\ominus}(N_2O_4)\} + RT\{2\ln a(NO_2) - \ln a(N_2O_4)\} \qquad (42)$$

■ Now use equation 41 to write an expression for $\Delta G_{\text{m}}^{\ominus}$ for reaction 19.

■ $$\Delta G_{\text{m}}^{\ominus} = 2\mu^{\ominus}(NO_2) - \mu^{\ominus}(N_2O_4) \qquad (43)$$

In other words, the first bracket in equation 42 can be replaced by $\Delta G_{\text{m}}^{\ominus}$. But what about the second bracket? This can be tidied up by recalling two important properties of logarithms that were revised during your study of Block 1 (again, refer to Section 1 of the AV Booklet if you are unsure about this):

$$a\ln x = \ln x^a$$

$$a\ln x + b\ln y = \ln x^a + \ln y^b$$

$$= \ln(x^a \times y^b)$$

■ Write the terms in the second bracket in this alternative way.

■ $$RT\{2\ln a(NO_2) - \ln a(N_2O_4)\} = RT(\ln\{a(NO_2)\}^2 + \ln\{a(N_2O_4)\}^{-1})$$

$$= RT\ln\frac{\{a(NO_2)\}^2}{a(N_2O_4)}$$

Substituting these expressions back into equation 42 gives

$$(dG/d\xi) = \Delta G_{\text{m}}^{\ominus} + RT\ln\frac{\{a(NO_2)\}^2}{a(N_2O_4)} \qquad (44)$$

The final step is to note that NO_2 and N_2O_4 are both gases. Assuming they behave as ideal gases, the definition of a in Case 2 leads to

$$a(NO_2) = p(NO_2)/p^{\ominus} \quad \text{and} \quad a(N_2O_4) = p(N_2O_4)/p^{\ominus}$$

where $p(NO_2)$ and $p(N_2O_4)$ are the partial pressures of NO_2 and N_2O_4, respectively, in the mixture. Then equation 44 becomes

$$(dG/d\xi) = \Delta G_{\text{m}}^{\ominus} + RT\ln\frac{\{p(NO_2)/p^{\ominus}\}^2}{p(N_2O_4)/p^{\ominus}} \qquad (45)$$

■ Does the form of equation 45 accord with the conclusions summarized at the end of Section 3?

Yes. Our measure of the 'tendency for reaction', $(dG/d\xi)$, is indeed determined both by the value of $\Delta G_m^\ominus$ (at the temperature T in question), and by the composition of the reaction mixture (the second term in equation 45).

To summarize: the technique outlined above is completely general; it can be used to derive an expression for $(dG/d\xi)$ for *any* reaction you care to write down. The key steps in the argument are collected together in Box 5.

Box 5 Deriving an expression for $(dG/d\xi)$

1 Write a balanced equation for the reaction:
$aA + bB + \ldots = pP + qQ + \ldots$

2 Use equation 35 to write an expression for $(dG/d\xi)$:
$$(dG/d\xi) = (p\mu_P + q\mu_Q + \ldots) - (a\mu_A + b\mu_B + \ldots) \qquad (35)$$

3 Use the general definition in equation 36 to expand the chemical potentials:
$$\mu = \mu^\ominus + RT\ln a \qquad (36)$$

4 Collect together the standard chemical potentials, $\mu^\ominus$, and call them $\Delta G_m^\ominus$.

5 Collect the activities together as one term with the form
$$RT\ln\left(\frac{a_P^p a_Q^q \ldots}{a_A^a a_B^b \ldots}\right)$$

6 Depending on the nature of the components (solids, liquids, gases, solutes, etc.) use the definitions in Cases 1–3 to replace the activities in the term $RT\ln(\ldots)$ with measurable quantities. In brief:

(a) For a liquid or solid: $a = 1$;

(b) For an ideal gas: $a = p/p^\ominus$ ($p^\ominus = 1$ bar);

(c) For a solute in an ideal solution: $a = c/c^\ominus$ ($c^\ominus = 1\ \text{mol dm}^{-3}$).

STUDY COMMENT Now try the following SAQ.

SAQ 7 Consider again the example in SAQ 5 (Section 3):

$$A(aq) = B(aq);\ \Delta G_m^\ominus(298.15\ \text{K}) = -5\ \text{kJ mol}^{-1}$$

Using the steps outlined in Box 5, derive an appropriate expression for $(dG/d\xi)$. (Assume that A and B are solutes in an ideal solution.) Hence confirm that the prediction you made then (when $c_B = 100c_A$) is in accord with the thermodynamic criterion, $(dG/d\xi) < 0$, for a spontaneous reaction.

We have achieved our objective. As the answer to SAQ 7 suggests, you are now in a position to *predict* the direction of spontaneous change in any chemical system, no matter what its composition – provided, of course, that the substances involved behave according to the definitions in Cases 1–3.*

* We have now introduced all the results necessary to calculate a plot like that in Figure 12. If you are interested, for the N_2O_4/NO_2 system the analysis starts with equation 23, $G = \mu(N_2O_4)n(N_2O_4) + \mu(NO_2)n(NO_2)$. Substituting for the chemical potentials (assuming ideal gas behaviour) and for the amounts in terms of the extent of reaction, leads finally to the following expression:

$$G = \mu^\ominus(N_2O_4) + \xi\Delta G_m^\ominus + RT\left\{(1-\xi)\ln\left(\frac{1-\xi}{1+\xi}\right) + 2\xi\ln\left(\frac{2\xi}{1+\xi}\right)\right\}$$

In drawing Figure 12, we arbitrarily set $\mu^\ominus(N_2O_4) = 0$.

4.6 The standard equilibrium constant

You may have recognized the *form* of one of the terms in equation 44. Suppose that the N_2O_4/NO_2 reaction system is at equilibrium: then $(dG/d\xi) = 0$, and the equation describing the system becomes

$$\textit{at equilibrium:} \quad 0 = \Delta G_m^{\ominus} + RT \ln \frac{\{a(NO_2)\}_e^2}{\{a(N_2O_4)\}_e} \qquad (46)$$

where the subscript e denotes the equilibrium values of the activities. Comparing this result with the thermodynamic relation in equation 17,

$$\Delta G_m^{\ominus} = -RT \ln K^{\ominus} \qquad (17)$$

enables us to give a formal definition of the **standard (or thermodynamic) equilibrium constant**, $K^{\ominus}$, which was introduced in Block 1. For the reaction in equation 19, it is just the ratio of the *activities at equilibrium* in equation 46:

$$K^{\ominus} = \frac{\{a(NO_2)\}_e^2}{\{a(N_2O_4)\}_e} \qquad (47)$$

■ Now write a comparable expression for the *dimensionless* equilibrium constant for reaction 19 in terms of partial pressures. (Look back at the answer to Exercise 1 if need be.)

■ For the reaction in equation 19, $K_p = \{p(NO_2)\}_e^2/\{p(N_2O_4)\}_e$. Dividing each partial pressure in this expression by the standard pressure $p^{\ominus}$ gives the following dimensionless ratio:

$$\frac{\{p(NO_2)/p^{\ominus}\}_e^2}{\{p(N_2O_4)/p^{\ominus}\}_e} \qquad (48)$$

In Block 1 (and hence in tackling the problem in Exercise 1), we identified this dimensionless ratio with the standard equilibrium constant $K^{\ominus}$, given by equation 47 in this case. It should now be clear that this identification does indeed carry with it the assumption that the gases involved behave ideally (when $a = p/p^{\ominus}$, and the two expressions become identical). As we suggested in Block 1, it turns out that this assumption breaks down seriously only under extreme conditions, and we shall continue to use it here.

Look again at the expressions in equations 44 and 47. Can you now spot a rather simple way of circumventing steps 2–5 in Box 5?

The ratio of *non-equilibrium* activities in equation 44 is of exactly the same form as the equilibrium constant $K^{\ominus}$ for the reaction. For this reason, it is often called the **reaction quotient**, and given the symbol Q: for a general 'alphabetical' reaction equation,

$$K^{\ominus} = \left(\frac{a_P^p a_Q^q \ldots}{a_A^a a_B^b \ldots}\right) \textit{at equilibrium} \qquad (49)$$

$$Q = \left(\frac{a_P^p a_Q^q \ldots}{a_A^a a_B^b \ldots}\right) \textit{under any arbitrary conditions} \qquad (50)$$

With this definition of Q, equation 44, for example, becomes:

$$(dG/d\xi) = \Delta G_m^{\ominus} + RT \ln Q \qquad (51)$$

Thus, an expression for $(dG/d\xi)$ *can be written down by simply inspecting the stoichiometry of the balanced reaction equation.* This simple result represents the culmination of our quest for a true thermodynamic measure of 'reaction tendency'. It also suggests another, particularly illuminating, way of interpreting that criterion – as shown in Figure 14 (cf. Figure 13).

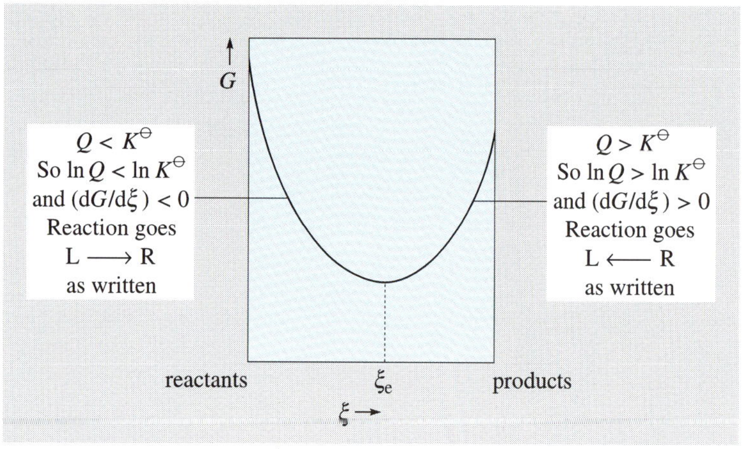

Figure 14 The direction of spontaneous change is determined by the *sign* of the slope of the 'Gibbs landscape'. For *any* reaction,

$$\begin{aligned}(dG/d\xi) &= \Delta G_m^\ominus + RT\ln Q\\ &= -RT\ln K^\ominus + RT\ln Q\\ &= RT\ln(Q/K^\ominus)\end{aligned}$$

Thus, the direction of change is effectively determined by the relative sizes of Q and $K^\ominus$.

STUDY COMMENT The following SAQ provides an opportunity to check that you can use the key points developed so far in this Section. Don't miss it out!

SAQ 8 Consider the cell discussed in Section 3, as represented by the following cell diagram:

$$\text{Pt, } H_2(g)|H^+(aq)|Ag^+(aq)|Ag(s)$$

with the implied cell reaction:

$$\tfrac{1}{2}H_2(g) + Ag^+(aq) = H^+(aq) + Ag(s) \qquad (16)$$

With the pressure of hydrogen gas held constant at 1 bar and at a temperature of 298.15 K, the following experimental observations were made:

$c(H^+)$/mol dm^{-3}	$c(Ag^+)$/mol dm^{-3}	Cell emf
(i) 10^{-5}	10^{-1}	$E > 0$
(ii) 10^{-5}	10^{-21}	$E < 0$

(a) Use the steps in Box 5 to write an expression for $(dG/d\xi)$ for the reaction in equation 16. Now write an expression for the reaction quotient Q: check that this confirms the validity of the shortcut implied by equation 51.

(b) Given that $\Delta G_m^\ominus = -77.1$ kJ mol^{-1} for reaction 16 at 298.15 K, calculate the value of $(dG/d\xi)$ for the two sets of conditions, (i) and (ii), above. What assumptions are inherent in these calculations?

(c) Are your results in agreement with the observed behaviour of this system, as reflected by the signs of the cell emf recorded above?

4.7 Electrolyte solutions: 'deviations' from ideality

As noted in the answer to SAQ 8, translating activities into molar concentrations involves the assumption that the solutes in question behave ideally. The final step in our thermodynamic analysis is to explore the validity of this assumption. To that end, we start by considering a deceptively simple example – the solubility of silver chloride, AgCl.

4.7.1 The solubility of silver chloride: a problem

Silver chloride is an ionic compound. Like a variety of other such compounds – familiar examples include many hydroxides and sulfides – it is only 'sparingly soluble' in water. Nevertheless, when AgCl is shaken up with water, some of the solid does dissolve, forming a **saturated solution**: the equilibrium between the solid and its aqueous ions can be represented as follows:

$$AgCl(s) = Ag^+(aq) + Cl^-(aq) \qquad (52)$$

In the chemical literature, such equilibria are commonly characterized by quoting the value of the **solubility product**, denoted K_{sp}. For AgCl, for example, at 298.15 K,

$$K_{sp} = c(Ag^+)c(Cl^-) \quad (53)$$
$$= 1.74 \times 10^{-10}\ mol^2\ dm^{-6}$$

- On this basis, what is the solubility, s, of AgCl in water at 298.15 K?
- According to equation 52, $s = c(Ag^+) = c(Cl^-)$, so $K_{sp} = s^2$ (equation 53), and $s = (K_{sp})^{1/2} = 1.32 \times 10^{-5}\ mol\ dm^{-3}$

As far as the solubility of AgCl in *pure* water is concerned, that is the end of the matter. But when AgCl is dissolved in an aqueous solution of a *different* electrolyte – magnesium sulfate ($MgSO_4$), for example – the situation changes markedly. As Figure 15 shows, the solubility of AgCl is now different from that in pure water, and *dependent* on the concentration of the 'foreign' electrolyte: in this case, the effect is quite substantial.

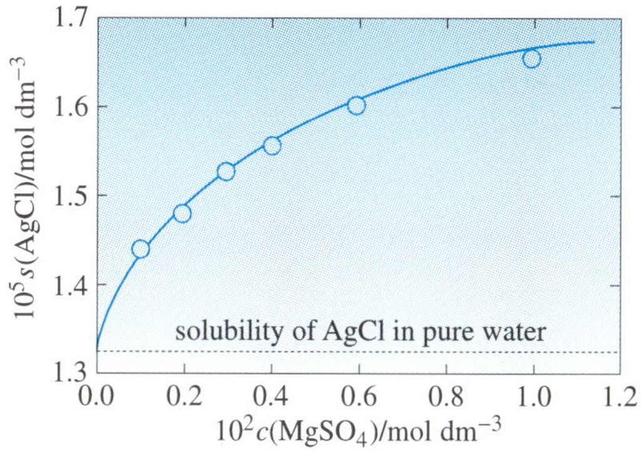

Figure 15 The solubility of AgCl in aqueous solutions of $MgSO_4$ of various concentrations (at 298.15 K).

- What does this suggest about the value of K_{sp} as defined by equation 53?
- If the solubility of AgCl changes, then so too must the concentrations of $Ag^+(aq)$ and $Cl^-(aq)$ change. K_{sp} does not appear to hold constant.

The problem highlighted by these experimental results can be defined more precisely by writing an expression for the *standard* equilibrium constant of reaction 52. Given the context, this is called the **standard solubility product**, and given the symbol $K_{sp}^{\ominus}$; from equation 49,

$$K_{sp}^{\ominus} = \frac{a(Ag^+)a(Cl^-)}{a(AgCl)}$$

Since the activity of the solid AgCl is unity, this becomes,

$$K_{sp}^{\ominus} = a(Ag^+)a(Cl^-) \quad (54)$$

- Considering the case of AgCl in pure water, compare the expressions in equations 53 and 54. What assumption must be made if $K_{sp}^{\ominus}$ is identified with the numerical magnitude of K_{sp}?
- The assumption of ideal behaviour – comparable with that involved in identifying $K^{\ominus}$ with the numerical magnitude of K_p for gaseous reactions (Section 4.6). In the present context, inserting in equation 54 the definition of activity for a solute in an ideal solution, $a = c/c^{\ominus}$ (Section 4.4, Case 3), leads to:

$$K_{sp}^{\ominus} = \{c(Ag^+)/c^{\ominus}\}\{c(Cl^-)/c^{\ominus}\} = K_{sp}/(c^{\ominus})^2 = 1.74 \times 10^{-10}$$

This relationship between $K_{sp}^{\ominus}$ and K_{sp} holds for AgCl in pure water. But for AgCl in aqueous $MgSO_4$, it cannot be true! Like any *standard* equilibrium constant, the value of $K_{sp}^{\ominus}$ depends only on temperature: it must *be* constant at a given temperature. Yet the value of K_{sp}, as defined by equation 53, is *not* constant. The most obvious explanation is that the assumption of ideal behaviour is no longer tenable. Or to put it more directly, in writing the standard equilibrium constant for reaction 52 it is not possible to use ionic concentrations in place of activities, that is,

$$K_{sp}^{\ominus} = a(Ag^+)a(Cl^-) \neq \{c(Ag^+)/c^{\ominus}\}\{c(Cl^-)/c^{\ominus}\}$$

Experiment shows that this type of **'deviation' from ideal behaviour** is particularly marked for electrolyte solutions, where the solute is dissociated into ions. Such 'deviations from ideality' are handled in a formal way by modifying the definition of activity in Case 3; *for a solute in a 'real' solution* it then reads

$$a = \gamma c / c^{\ominus} \qquad (55)$$

where γ (Greek 'gamma') denotes the **activity coefficient**. It is a pure number that it is taken to be a 'measure' of the deviations from ideality. Or to put it another way, $\gamma = 1$ implies ideal behaviour: the more different from 1 it becomes, the more 'non-ideal' is the solution. It is through the activity coefficient that a solute 'responds' to changes in the composition of its environment – a response that can lead to behaviour like that in Figure 15.

- Given the general definition of activity in equation 36 ($\mu - \mu^{\ominus} = RT \ln a$), when does the chemical potential of a 'real' solute have its standard value, $\mu^{\ominus}$?
- According to equation 55, $a = 1$ (that is, $\mu = \mu^{\ominus}$) when $\gamma c = c^{\ominus} = 1 \text{ mol dm}^{-3}$.

Although this is true, the solute is *not* then in its standard state: it merely has the same chemical potential as it does in the standard state. The definition of the standard state itself is now quite subtle. It goes as follows:

The standard state of a solute in solution is the *hypothetical* state of unit molarity ($c = c^{\ominus} = 1 \text{ mol dm}^{-3}$) where all the molecular interactions leading to deviations from ideality have been eliminated (that is, $\gamma = 1$).

If the significance of this hypothetical state escapes you, bear with us! It should become clearer in Section 4.7.4, where you will see that, *in practice*, the properties of the standard state *can* be obtained – by measuring values at various concentrations, and then finding the 'limiting' value as the solution becomes progressively more and more dilute.

4.7.2 Ionic activity coefficients

To return to the solubility of AgCl: using the modified definition of activity above, equation 54 can be rewritten as

$$K_{\text{sp}}^{\ominus} = (a_+)(a_-) = (\gamma_+ c_+ / c^{\ominus})(\gamma_- c_- / c^{\ominus}) \qquad (56)$$

where the simplified subscripts + and – have been used to represent Ag^+ and Cl^-, respectively.

Suppose now that you wanted to test whether the expression in equation 56 does indeed hold constant (under the conditions in Figure 15, for example): obviously, you would need to determine the values of γ_+ and γ_-. But a solution of AgCl (or any other electrolyte for that matter) contains *both* cations *and* anions: perhaps you can see intuitively that there is no way of 'disentangling' the product $\gamma_+\gamma_-$ experimentally, and of assigning one part to the cations and another to the anions. In other words, individual ionic activity coefficients *cannot* be measured.

The way around this difficulty is to define the activity of an ion only in the presence of 'counter ions' (that is, ions of opposite charge). In practice, this is achieved by introducing a **mean ionic activity coefficient**, $\gamma_\pm$. For a 1 : 1 electrolyte, such as AgCl or $MgSO_4$, it is defined as follows:

$$\gamma_\pm = (\gamma_+ \gamma_-)^{1/2} \qquad (57)$$

Analogous definitions for other common 'types' of electrolyte are collected in Table 3.

Table 3 Definition of the mean ionic activity coefficient, $\gamma_\pm$, for electrolytes of the general type M_pX_q.[a]

Formula	Examples	p	q	Definition of $\gamma_\pm$
MX	AgCl, $MgSO_4$	1	1	$(\gamma_+\gamma_-)^{1/2}$
MX_2	$MgCl_2$, $Ca(NO_3)_2$	1	2	$(\gamma_+\gamma_-^{2})^{1/3}$
M_2X	K_2SO_4	2	1	$(\gamma_+^{2}\gamma_-)^{1/3}$
MX_3	$Fe(OH)_3$	1	3	$(\gamma_+\gamma_-^{3})^{1/4}$
M_2X_3	$Al_2(SO_4)_3$	2	3	$(\gamma_+^{2}\gamma_-^{3})^{1/5}$
⋮		⋮	⋮	
M_pX_q		p	q	$(\gamma_+^{p}\gamma_-^{q})^{1/n}$; $n = p + q$

[a] M represents the cation, X the anion.

■ Use this definition to rewrite equation 56 in terms of $K_{sp}^{\ominus}$, K_{sp} (equation 53) and $\gamma_\pm$.

■ $$K_{sp}^{\ominus} = (\gamma_+\gamma_-)(c_+/c^{\ominus})(c_-/c^{\ominus})$$
$$= \gamma_\pm^{2}\, K_{sp}/(c^{\ominus})^2 \qquad (58)$$

The question now arises as to whether $\gamma_\pm$ can be determined experimentally. Indeed it can. Equation 58 holds the clue to one technique, which forms the basis of SAQ 9. Later on you will meet an alternative, and in some ways more general, technique – based on the use of an appropriate electrochemical cell.

SAQ 9 As noted in Section 4.7.1, $K_{sp}^{\ominus}$ for AgCl is just the standard equilibrium constant for reaction 52:

$$AgCl(s) = Ag^+(aq) + Cl^-(aq) \qquad (52)$$

(a) Use information from the S342 *Data Book* to calculate $K_{sp}^{\ominus}$ for AgCl at 298.15 K.

(b) Two of the experimental points from Figure 15 are given below (s is the solubility of AgCl in a $MgSO_4$ solution of concentration c). Calculate the corresponding values of $\gamma_\pm$ for AgCl.

	(i)	(ii)
$10^5 s$/mol dm^{-3}	1.48	1.60
$10^2 c$/mol dm^{-3}	0.20	0.60

The answers to part (b) of SAQ 9 highlight an important general point: evidently, the mean ionic activity coefficient of AgCl depends on the concentration, not only of its 'own' ions, as it were, but also of the ions of 'foreign' electrolyte ($MgSO_4$) in the solution. Why this should be so is taken up below.

4.7.3 A picture of ions in solution: the Debye–Hückel theory

Experimental values of mean ionic activity coefficients for various (relatively soluble) electrolytes are collected in Table 4: evidently, the value of $\gamma_\pm$ for a given electrolyte depends strongly on the concentration of the solution. Although the detailed trends in Table 4 are far from simple, broadly speaking the more concentrated the solution, the greater the deviations from ideality – that is, the more different from unity is $\gamma_\pm$. And the effect appears to be more pronounced if one (or both) of the ions is highly charged. But what is the underlying reason for this behaviour?

Table 4 Mean ionic activity coefficients, $\gamma_{\pm}$, for various electrolytes as a function of their concentration c (in aqueous solution at 298.15 K).

c/mol dm^{-3}	HCl	NaOH	$CaCl_2$	H_2SO_4	$ZnSO_4$	$Al_2(SO_4)_3$
0.001	0.966	–	0.888	0.830	0.734	–
0.01	0.904	–	0.732	0.544	0.387	–
0.10	0.796	0.766	0.524	0.265	0.148	(0.035)
0.50	0.758	0.693	0.510	0.155	0.063	0.014
1.00	0.809	0.679	0.725	0.130	0.043	0.018
2.00	1.010	0.700	1.554	0.124	0.035	–

The clue lies in the fact that ions are charged: oppositely charged ions attract one another. Although the solution is neutral overall, this suggests that cations and anions are unlikely to be distributed completely at random, and therefore *uniformly*, throughout the solution. Rather, we would expect to find a predominance of anions in the vicinity of cations, and vice versa (Figure 16a).*

Suppose now that Figure 16a is a snapshot of the ionic distribution at a given instant in time. Ions are constantly on the move, so different snapshots would show different distributions. Nevertheless, scrutiny of a sufficient number of pictures would allow recognition of a certain time-average distribution, in which any given ion would be surrounded by a 'spherical haze' of opposite charge – the so-called **ionic atmosphere** of the ion (Figure 16b).

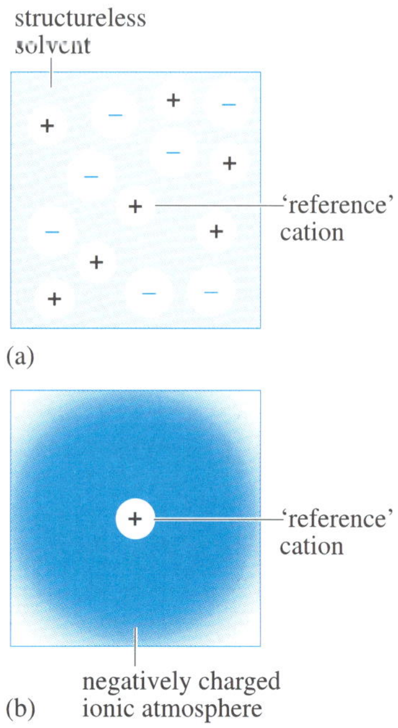

Figure 16 A representation of the 'molecular' picture that underlies the Debye–Hückel theory.

This simple picture of an ionic solution underlies the interpretation of ionic activity coefficients, according to a molecular model first propounded by P. Debye and E. Hückel in 1923. Its genius was its simplicity, for the theory argues that the deviations from ideality (contained in $\gamma_{\pm}$) are due solely to electrostatic interactions among the ions present – interactions that lead to the sort of average distribution indicated in Figure 16b.

To spell this out a little, it helps to start with an expression for the total chemical potential of the ions (cations *plus* anions) in a *real* – and hence 'non-ideal' – electrolyte solution. For a 1 : 1 electrolyte, where $c_+ = c_- = c$ and $\gamma_{\pm} = (\gamma_+\gamma_-)^{1/2}$, this can be written:

$$\begin{aligned}\mu_{\text{real}} &= \mu_+ + \mu_- \\ &= (\mu_+^{\ominus} + RT\ln a_+) + (\mu_-^{\ominus} + RT\ln a_-) \\ &= (\mu_+^{\ominus} + \mu_-^{\ominus}) + RT\ln(\gamma_+ c_+/c^{\ominus}) + RT\ln(\gamma_- c_-/c^{\ominus}) \\ &= (\mu_+^{\ominus} + \mu_-^{\ominus}) + RT\ln(c/c^{\ominus}) + RT\ln(c/c^{\ominus}) + RT\ln(\gamma_+\gamma_-) \\ &= (\mu_+^{\ominus} + \mu_-^{\ominus}) + 2RT\ln(c/c^{\ominus}) + RT\ln(\gamma_{\pm})^2\end{aligned}$$

$$\text{So } \mu_{\text{real}} = (\mu_+^{\ominus} + \mu_-^{\ominus}) + 2RT\ln(c/c^{\ominus}) + 2RT\ln\gamma_{\pm} \quad (59)$$

By contrast, in an *ideal* solution, $\gamma_{\pm} = 1$, and equation 59 becomes

$$\mu_{\text{id}} = (\mu_+^{\ominus} + \mu_-^{\ominus}) + 2RT\ln(c/c^{\ominus}) \quad (60)$$

Now look again at Figure 16a and do a thought experiment. Imagine a solution in which all the ions have the positions shown here, but in which the charges on the ions are 'switched off', so that the electrostatic interactions between them cease (Figure 17a). If the deviations from ideality are due to these interactions alone, then this hypothetical situation must correspond to an ideal solution, for which the chemical potential is given by equation 60. The determination of $\gamma_{\pm}$ then reduces to finding the *change* in chemical potential when the charges on the ions are returned to their true values, while their average distribution is held constant (Figure 17b).

* As you may recall from the Second Level Inorganic Course, an electrolyte solution actually consists of *hydrated* ions *and* water molecules. In the simple model considered here, the water molecules (including those 'attached' to the ions, as it were) are looked upon as a continuous, structureless medium. Because water molecules are polar, however, their presence *reduces* the electrostatic forces between the ions – by a factor that is taken to be equivalent to the *dielectric constant*, or relative permittivity, of bulk water. This factor (roughly 80, relative to a vacuum) must be taken into account in the quantitative development of the model.

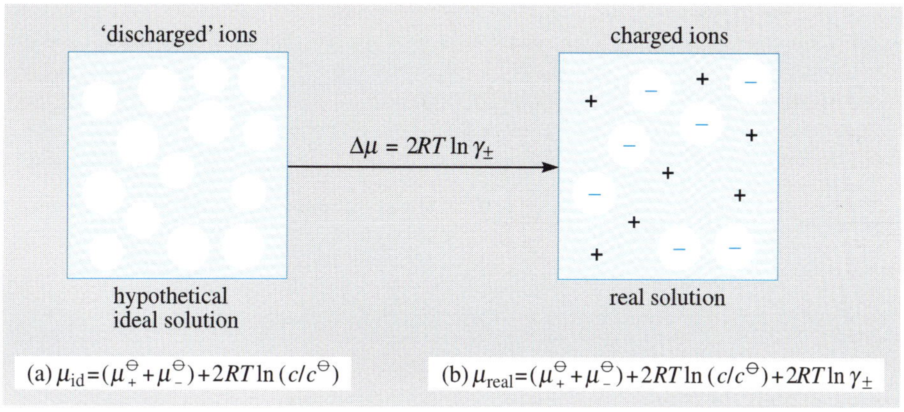

Figure 17 According to the Debye–Hückel theory, $\gamma_{\pm}$ may be calculated from the chemical potential change, $\Delta\mu = \mu_{real} - \mu_{id} = 2RT \ln \gamma_{\pm}$, in going from a hypothetical ideal electrolyte solution (a), in which ion-ion interactions do not operate, to a real solution (b), in which these interactions do operate.

The detailed steps whereby the Debye–Hückel theory does just that require a background – in both mathematics and electrostatics – beyond that assumed in this Course: here we simply quote the final result of that analysis,

$$\log \gamma_{\pm} = -A|z_+z_-|\,I^{1/2} \qquad (61)$$

where log represents the logarithm to the base ten.

To define the terms in equation 61: z_+ and z_- are just the charges on the positive and negative ions, respectively. The *modulus sign* | | means the *magnitude* of the product z_+z_-, without its sign. Thus for AgCl, for instance, $z_+ = +1$ and $z_- = -1$, whence $z_+z_- = -1$, but $|z_+z_-| = 1$.

The term I is a little more involved. The Debye–Hückel theory argues that the level of electrostatic interactions is a function, not only of the *concentrations* of *all* the ions in solution (as noted at the end of Section 4.7.2), but also of the *charges* on those ions. This certainly ties in with our comment about the experimental values of $\gamma_{\pm}$ in Table 4. A measure of the overall effect is the **ionic strength**, I, which is defined as follows:

$$I = \tfrac{1}{2}\sum(c_i/c^{\ominus})z_i^2 \qquad (62)$$

where c_i is the concentration of ion i of charge z_i, and $c^{\ominus}$ is again the standard concentration (= 1 mol dm^{-3}). Thus, to determine the ionic strength of a solution, all the individual values of $(cz^2/c^{\ominus})$ must be added together, one term for *each* species of ion present in the solution. Notice that with concentrations expressed in mol dm^{-3}, I is dimensionless.

- Calculate the ionic strength in a saturated solution of AgCl in pure water at 298.15 K (when $s = 1.32 \times 10^{-5}$ mol dm^{-3}).

- Here, $c(Ag^+) = c(Cl^-) = 1.32 \times 10^{-5}$ mol dm^{-3} (Section 4.7.1), so

  $I = \tfrac{1}{2}(\{c(Ag^+)/c^{\ominus}\}(+1)^2 + \{c(Cl^-)/c^{\ominus}\}(-1)^2) = 1.32 \times 10^{-5}$

Thus, for a solution of a single 1 : 1 electrolyte of *univalent* ions – such as AgCl or NaOH – the ionic strength is just the concentration of the electrolyte (or strictly, the numerical magnitude of that concentration). But this is not always so. In general, the link between concentration and ionic strength depends on the valence types of the ions – even for a solution containing a single electrolyte. More importantly, in a solution containing *several* electrolytes, each ionic species makes its own contribution to the overall ionic strength (see SAQ 10, below).

Returning to equation 61, A is a constant, the value of which depends on certain physical properties of the solvent (notably its density and dielectric constant), and the temperature. For water at 298.15 K, $A = 0.51$, and equation 61 becomes

$$\log \gamma_{\pm} = -0.51|z_+z_-|\,I^{1/2} \qquad (63)$$

For reasons that will become apparent in the next Section, equation 63 is known as the **Debye–Hückel limiting law**.

SAQ 10 (a) Calculate the ionic strength in a saturated solution of AgCl in 0.002 molar $MgSO_4$, in which the solubility of AgCl is $s = 1.48 \times 10^{-5}$ mol dm^{-3}.

(b) According to Figure 15, the solubility of AgCl depends on the presence of 'foreign' ions in the solution. Does the *general* form of equation 63 tie in with this behaviour?

4.7.4 The strengths and limitations of the Debye–Hückel theory

To examine the implications of equation 63 more closely, consider Figure 18: it shows the predicted concentration-dependence of log $\gamma_\pm$ for a number of different electrolytes, compared with that found experimentally. Notice, first of all, that the experimental plots converge to a single point, log $\gamma_\pm = 0$, as the ionic strength tends to zero, as predicted by equation 63. In other words, $\gamma_\pm \rightarrow 1$ (that is, it tends to unity) as the solution becomes progressively more dilute. In this limit of zero ionic strength, the solution is described as being **infinitely dilute**: in this limit, the ionic interactions become negligible and the solution then behaves ideally (that is, $\gamma_\pm = 1$). We come back to this point in a moment.

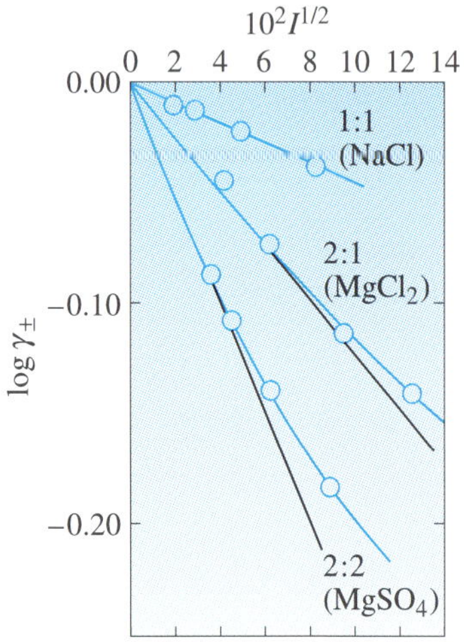

Figure 18 The observed (blue) dependence of the mean ionic activity coefficient on ionic strength for various valence types, compared with the predictions (black) of the Debye–Hückel limiting law.

- According to equation 63, a plot of log $\gamma_\pm$ against $I^{1/2}$ should be linear. What property of the electrolyte should determine the slope?
- The 'valence type' of the electrolyte, that is, the charges z_+ and z_- on its ions.

Again, these predictions are borne out by the comparison in Figure 18 – up to a point, at least. However, a divergence between theory and experiment is already apparent for the higher valence types in Figure 18: even for 1 : 1 electrolytes, it becomes overwhelming as soon as the comparison is carried to higher values of the ionic strength (Figure 19).

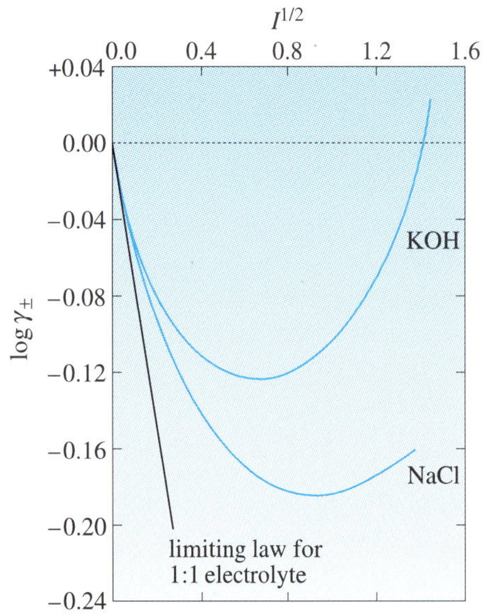

Figure 19 Even though NaCl and KOH are both 1:1 electrolytes containing univalent ions, their activity coefficients vary in different ways with ionic strength (or concentration) as soon as the comparison is taken to higher concentrations.

To summarize: it seems that equation 63 is an approximation, the validity of which is strictly restricted to very dilute solutions (less than 10^{-2} or 10^{-3} molar, depending on valence type – or 'slightly polluted water', as it has sometimes been termed!). In other words, as Figure 19 suggests, it provides a *limiting law* – a type of behaviour to which *all* electrolyte solutions tend as the concentration of ions becomes very low. In a sense, however, this very limitation is the key to one of the most important uses of equation 63.

As an example, consider again the solubility of silver chloride. According to the answer to SAQ 9(b), $K_{sp}^{\ominus}$ for AgCl can be written as follows:

$$K_{sp}^{\ominus} = \gamma_{\pm}^{2}(s/c^{\ominus})^2 \quad (64)$$

where s is the observed solubility of AgCl in solutions of $MgSO_4$ of various molarites, as in Figure 15.

Suppose now that you wanted to use the data in Figure 15 to *determine* the standard solubility product $K_{sp}^{\ominus}$. The first step is to take logarithms (to the base of 10) of equation 64:

$$\log K_{sp}^{\ominus} = \log (\gamma_{\pm})^2 + \log (s/c^{\ominus})^2 = 2 \log \gamma_{\pm} + 2 \log (s/c^{\ominus})$$

which can be divided by 2 and reorganized to give:

$$\log (s/c^{\ominus}) = \tfrac{1}{2} \log K_{sp}^{\ominus} - \log \gamma_{\pm} \quad (65)$$

Using the limiting law expression in equation 63 (with $|z_+z_-| = 1$ for AgCl) this becomes:

$$\log (s/c^{\ominus}) = \tfrac{1}{2} \log K_{sp}^{\ominus} + 0.51\, I^{1/2} \quad (66)$$

Figure 20 shows the data of Figure 15 replotted according to this equation. Given the discussion above, you should now be able to determine $K_{sp}^{\ominus}$ for yourself by doing the following SAQ.

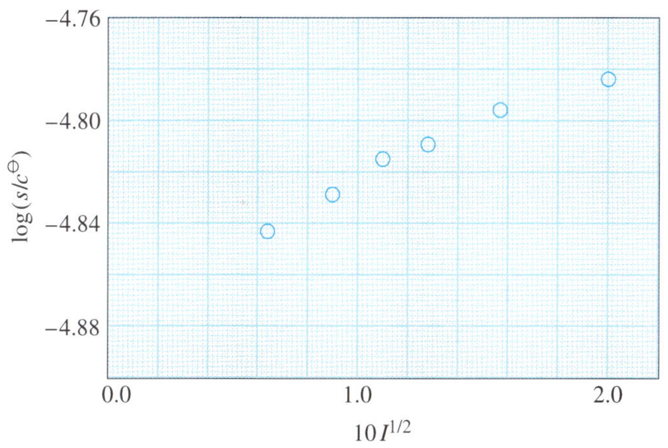

Figure 20 The solubility of AgCl as a function of ionic strength: the data of Figure 15 replotted according to equation 66.

SAQ 11 (a) Use an extrapolation based on equation 66 to estimate $K_{sp}^{\ominus}$ for AgCl (at 298.15 K) from the plot in Figure 20. [*Hint* The equation for a straight line can be written in general terms as $y = mx + c$. By comparing this expression with the one in equation 66, identify each of the quantities x, y, m and c in this case. That should give you a good idea of how to proceed.]

(b) To what extent does the plot in Figure 20 accord with our conclusions regarding the limitations of equation 63?

One final point. In Section 4.7.1 we defined the standard state of a solute in solution as the hypothetical state of unit molarity where all the molecular interactions leading to deviations from ideality have been eliminated. But according to the discussion above, in the limit of infinite dilution, *any* solute behaves ideally (that is, the activity coefficient becomes unity). For an ionic solution, the Debye–Hückel limiting law provides a concrete measure of ‘infinite dilution’, namely zero ionic strength. It thereby suggests an invaluable technique for determining *standard* thermodynamic quantities ($K_{sp}^{\ominus}$, for example): measure the required quantity as a function of ionic strength and then extrapolate to $I = 0$. You will see the importance of this technique in a rather different context in Section 7.2.

4.8 Summary of Section 4

1 The criterion for spontaneous change in a chemical system (at constant T and p) can be formulated more precisely as: $(\mathrm{d}G/\mathrm{d}\xi) < 0$. At equilibrium, $(\mathrm{d}G/\mathrm{d}\xi) = 0$.

2 A chemical reaction can be represented by a balanced equation. The corresponding expression for $(\mathrm{d}G/\mathrm{d}\xi)$, at a constant temperature T (and implicitly at constant pressure), can be written down by inspecting the stoichiometry of that equation:

$$(\mathrm{d}G/\mathrm{d}\xi) = \Delta G_\mathrm{m}^{\ominus}(T) + RT \ln Q \qquad (51)$$

(a) The value of $\Delta G_\mathrm{m}^{\ominus}(T)$ can be obtained by consulting a convenient source of thermodynamic data (such as the S342 *Data Book*): it is related to the standard equilibrium constant $K^{\ominus}(T)$ for the reaction as follows:

$$\Delta G_\mathrm{m}^{\ominus}(T) = -RT \ln K^{\ominus}(T)$$

where $K^{\ominus}(T)$ is the ratio of equilibrium activities appropriate to the balanced reaction equation:

$$K^{\ominus} = \left(\frac{a_\mathrm{P}^{p} a_\mathrm{Q}^{q} \ldots}{a_\mathrm{A}^{a} a_\mathrm{B}^{b} \ldots}\right) \text{ at equilibrium} \qquad (49)$$

(b) The reaction quotient Q is the analogous ratio of non-equilibrium activities, appropriate to some arbitrary composition of the reaction mixture.

3 The determination of $(\mathrm{d}G/\mathrm{d}\xi)$ from equation 51 depends on the definition of the activity a: this, in turn, depends on the nature of the substances involved. To summarize:

(a) For a liquid or solid: $a = 1$;

(b) For a gas: $a = p/p^{\ominus}$ ($p^{\ominus} = 1$ bar);

(c) For a solute in an 'ideal' solution: $a = c/c^{\ominus}$ ($c^{\ominus} = 1\ \mathrm{mol\ dm^{-3}}$);

(d) For a solute in a 'real' solution: $a = \gamma c/c^{\ominus}$, where γ is the activity coefficient.

4 For ionic solutions, the deviations from ideality can be attributed to electrostatic interactions between the ions, as measured by the mean ionic activity coefficient $\gamma_\pm$ for a given electrolyte.

5 In very dilute solutions, the Debye–Hückel limiting law (equation 63) allows the calculation of $\gamma_\pm$ from the ionic strength I of the solution:

$$\log \gamma_\pm = -0.51|z_+z_-|I^{1/2} \qquad (63)$$

where

$$I = \tfrac{1}{2}\Sigma(c_i/c^{\ominus})z_i^2 \qquad (62)$$

Alternatively, it provides a technique for extrapolating measured values of a thermodynamic quantity to 'infinite dilution', and thereby determining the standard value of that quantity.

SAQ 12 (a) Barium sulphate is a sparingly soluble salt:

$$BaSO_4(s) = Ba^{2+}(aq) + SO_4^{2-}(aq)$$

Use information from the S342 *Data Book* to calculate $K_\mathrm{sp}^{\ominus}$ for $BaSO_4$ at 298.15 K.

(b) Use your value of $K_\mathrm{sp}^{\ominus}$ to calculate the solubility of $BaSO_4$ in pure water at 298.15 K. State any assumptions involved in your calculation.

(c) Use the Debye–Hückel limiting law to calculate $\gamma_\pm$ of $BaSO_4$ in a solution of $NaNO_3$ of concentration 0.01 mol dm^{-3}, and hence estimate the solubility of $BaSO_4$ in this solution. Again state any assumptions involved in your calculation.

STUDY COMMENT The following Exercise provides an opportunity for you to draw together and apply most of the ideas developed in Section 4. It also sets the scene for our return to electrochemical cells.

EXERCISE 2

Consider again the Daniell cell (SAQ 4, Section 2.5):

$Zn(s)|Zn^{2+}(aq)|Cu^{2+}(aq)|Cu(s)$

with the implied cell reaction:

$$Zn(s) + Cu^{2+}(aq) = Zn^{2+}(aq) + Cu(s) \quad (67)$$

$$\Delta G_m^{\ominus}(298.15\,K) = -212.6\,kJ\,mol^{-1}$$

(a) Write expressions for $K^{\ominus}$ and for $(dG/d\xi)$ for the reaction in equation 67. Can you spot a problem with the expressions you obtain?

(b) If the activities of $Zn^{2+}(aq)$ and $Cu^{2+}(aq)$ in the separate half-cells are each unity, what is the corresponding value of $(dG/d\xi)$?

(c) Suppose now that the electrolytes in the two half-cells are $ZnSO_4(aq)$ and $CuSO_4(aq)$, respectively, with the following compositions:

	c/mol dm^{-3}	$\gamma_{\pm}$
$ZnSO_4(aq)$	0.1	0.148
$CuSO_4(aq)$	1.0	0.043

Calculate the value of $(dG/d\xi)$ under these conditions.

(d) Calculate $K^{\ominus}$, and hence the ratio of activities at equilibrium. Under what conditions would you expect the spontaneous cell reaction to be *reversed*, such that the copper electrode became the anode? Is this likely in practice? (You may find it useful to refer back to the answer to SAQ 8.)

(e) What do you conclude about the circumstances under which it is reasonable to take the sign of $\Delta G_m^{\ominus}$ *alone* as the criterion for a spontaneous reaction?

5 AN ALTERNATIVE MEASURE OF 'REACTION TENDENCY'

Armed with the thermodynamic ideas developed in the previous Section, we are now in a position to return to our study of electrochemical cells, and to the problems and questions raised in Sections 2 and 3.

As you should have found in answering SAQ 8 and Exercise 2, the analysis that culminated in equation 51,

$$(dG/d\xi) = \Delta G_m^{\ominus}(T) + RT\ln Q \quad (51)$$

has already gone a long way toward rationalizing the observed behaviour of electrochemical cells. In particular, it certainly resolves the central problem identified in Section 3: the direction of spontaneous change in a given cell (that is, the sign of $(dG/d\xi)$ for the underlying cell reaction) evidently depends on a balance between the two terms on the right-hand side of equation 51. Although the value of $\Delta G_m^{\ominus}$ is fixed (at a particular temperature), the second term is open to manipulation: this is sometimes (SAQ 8), but by no means always (Exercise 2), sufficient to tip the scales in favour of the reverse reaction.

Equally, the results summarized above lend support to our hypothesis that the emf, E, of a cell is an alternative measure of reaction tendency. According to the convention introduced in Section 2.3.2 (Box 3), the *sign* of E, like the sign of $(dG/d\xi)$, is related to the direction of spontaneous change: at equilibrium, both quantities become zero.

All that remains, then, is to establish a formal connection between the two quantities. The link is provided by the answer to a question that has been given scant attention so far (Question 1 in Section 2.1): How much work can be obtained from a particular reaction, and how does this depend on the way in which that reaction is carried out?

5.1 Work from an electrochemical cell

To tackle the question posed above requires a brief review of the ideas on chemical energy developed in the Second Level Inorganic Course. With this in mind, consider again the reaction with which we began our study of electrochemical cells:

$$Cu(s) + 2Ag^+(aq) = Cu^{2+}(aq) + 2Ag(s) \quad (8)$$

Now, it is found by experiment that the enthalpy change for reaction of 10^{-3} mol (0.063 g) of copper, according to this equation, has the following value (at 298.15 K and 1 bar pressure):

$$\Delta H = -146.4\ \text{J}$$

■ The reaction is exothermic: it releases energy. Do you recall the two *distinct* ways in which this energy could be transferred to the surroundings?

■ In general, energy can be transferred either as **heat**, q, or as **work**, w.

In the Second Level Inorganic Course (as here) we were mainly interested in **electrical work**, w_{el}: this led us to formulate the **first law of thermodynamics** (the principle of energy conservation) as follows:*

$$\Delta H = q + w_{el} \quad \text{(at constant } p) \quad (68)$$

where q is the heat *added to* the system, and w_{el} is the electrical work *done on* the system – that is, positive values of q and w_{el} imply that energy is transferred *to* the system from the surroundings, and vice versa for negative values.

■ If the reaction cited above takes place in an open beaker (at a constant temperature of 298.15 K), how is the energy transfer apportioned between heat and work?

■ Under these conditions, $w_{el} = 0$ and $\Delta H = q = -146.4$ J: all the energy released is *lost* from the system, and heats up the surroundings.

Shifting attention away from the system itself, and concentrating on what happens to the *surroundings*, then it seems that *minus* ΔH is a measure of the **maximum heat** (146.4 J in this case) that we can get *from* a reaction. Indeed, most of the energy we use today relies, ultimately, on this conversion of 'chemical energy' into heat.

Suppose now that the system in equation 8 is set up in an electrochemical cell: the underlying cell reaction can then be made to 'do' electrical work. We could, for example, include a small electric motor in the external circuit (Figure 21), and then allow the same change as before (that is, dissolution of 10^{-3} mol of copper from the anode) to take place. But what is the *maximum* work we can hope to obtain in this way?

* As noted in the Second Level Inorganic Course, equation 68 is a special and restricted case of the first law of thermodynamics – restricted in that it applies only to changes at constant p, and that it considers explicitly only electrical work. Once again, these conditions are satisfied by the examples you will meet in this Course. (You may recall that a change at constant p will almost always be accompanied by expansion work; but the definition of enthalpy effectively includes this term, so there is no need to take it into account explicitly.)

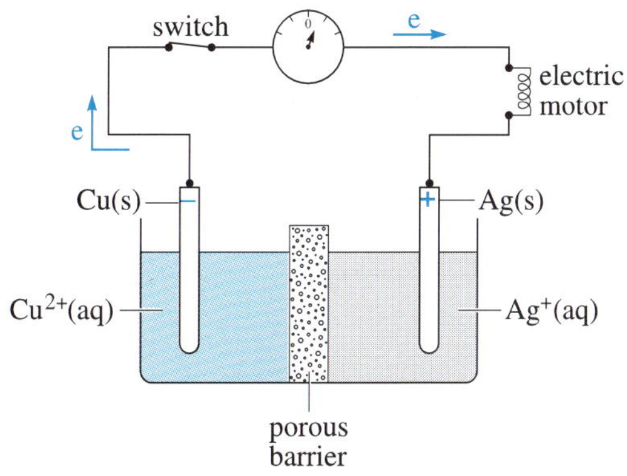

Figure 21 Work from an electrochemical cell: the electrical load in the external circuit (e.g. an electric motor) is simply represented as an equivalent resistance. (The porous barrier has the same role as the salt bridge.)

At first sight, there is a beguilingly simple answer to this question. Look back at the statement of the first law in equation 68. If we can somehow ensure that the heat transfer $q = 0$, then $\Delta H = w_{el}$ and *all* the energy released by this exothermic reaction will be transformed into work.

Unfortunately, this has simply never been observed – not with this, nor with countless other reactions. Rather, accumulated experience suggests that if a reaction takes place at constant T and p, then the heat transfer *q cannot, in general, be reduced to zero*. The best we can do is to reduce q to a minimum – and this only when the reaction is carried out **reversibly**.

The meaning of this term is best explained by reference to Figure 6 (Section 2.2.1). Thus, to do the reaction in a reversible way it is necessary to ensure that the cell is, *at all times*, only marginally 'out of balance'. The reaction is then allowed to proceed in a series of infinitesimal steps, *during each of which a vanishing small current is drawn from the cell*, until ultimately the desired change (reaction of 10^{-3} mol of Cu according to equation 8, for example) has taken place. In theory, this can be achieved with the help of an external 'balancing' potential, as described in Section 2.2.1. In practice, one can envisage using an electric motor with an extremely high equivalent resistance, which would be driven at an infinitesimal rate by the minute current.

Under these, admittedly rather bizarre, conditions, the heat transfer can be labelled q_{rev}, and equation 68 becomes,

$$\Delta H = q_{rev} + w_{el,\,max} \quad \text{(reversible change; constant } T \text{ and } p\text{)} \qquad (69)$$

where $w_{el,\,max}$ is now the **maximum electrical work** done.

Equation 69 can be taken a stage further by making use of another thermodynamic relation introduced in the Second Level Inorganic Course. *For a reversible change* (at constant T and p), the heat transfer q_{rev} is *defined* as follows:

$$q_{rev} = T\Delta S \text{ (by definition)} \qquad (70)$$

where ΔS is the corresponding entropy change (for the process cited above, in this case). Using this relation, equation 69 becomes

$$\Delta H = T\Delta S + w_{el,\,max}$$

or

$$\Delta H - T\Delta S = w_{el,\,max} \qquad (71)$$

- Can you now see the thermodynamic quantity that *does* provide a measure of maximum work?

- For a change at constant T and p, $\Delta G = \Delta H - T\Delta S$ so equation 71 becomes

$$\Delta G = w_{el,\,max} \qquad (72)$$

Again switching our attention to the *surroundings*, equation 72 implies that it is $-\Delta G$, not $-\Delta H$, that provides a theoretical limit to the amount of work *obtainable* from a spontaneous reaction (where $\Delta G < 0$). Indeed, this is the reason for the alternative name for the Gibbs function – the 'free' energy: it represents the chemical energy that is free, or 'available', for conversion into useful work.

At this point, we should again stress that equation 72 holds *only* under the very special (and rather impractical!) conditions outlined above. For a real change (as when an appreciable current is drawn from the cell) the work output will always be less than the maximum, the difference appearing as an additional heat loss to the surroundings (in accordance with the first law of thermodynamics).

SAQ 13 The entropy change for the reaction discussed in this Section has the following value, at 298.15 K and 1 bar pressure:

$$\Delta S = -193.2 \times 10^{-3}\,\mathrm{J\,K^{-1}}$$

What are the corresponding values of q_{rev} and $w_{\mathrm{el,\,max}}$? Is your value for the latter in accord with the sign convention for work done?

One final point. Equation 72 relates to a finite change in the system (see the answer to SAQ 13): an entirely equivalent expression can be written for an *infinitesimal* change:

$$\mathrm{d}G = w_{\mathrm{el,\,max}} \qquad (73)$$

Equation 73 is the vital first step in establishing a link between $(\mathrm{d}G/\mathrm{d}\xi)$ and the corresponding cell emf, E. Not surprisingly, the next step is to relate $w_{\mathrm{el,\,max}}$ to E. To this end, we concentrate now on the flow of electrons through the *external* circuit of a cell like the one in Figure 21.

5.2 Electrical work and the cell emf

For a self-driving electrochemical cell (like the one in Figure 21), electrons flow spontaneously through the external circuit from the negative electrode (anode) to the positive electrode (cathode). One way of explaining why this happens is to say that the electrons move to the positive electrode because their **electrical energy** is lower at that terminal. This is analogous to saying that an object will fall to the Earth when released, because the object's gravitational energy is thereby reduced.

Put more precisely, it can be shown that the change in electrical energy when a charge moves from one position to another is determined by the potential difference between those two positions: specifically,

$$\text{electrical energy difference} = \text{potential difference} \times \text{charge} \qquad (74)$$

- Does the right-hand side of equation 74 produce something with the unit of energy?

- Yes. In SI, the units of potential difference and charge are volt (V) and coulomb (C), respectively. Then $\mathrm{V} \times \mathrm{C} = (\mathrm{J\,A^{-1}\,s^{-1}}) \times (\mathrm{A\,s}) = \mathrm{J}$, the unit of energy.

To begin with, suppose that an infinitesimal amount of charge, $\mathrm{d}Q$ say, flows from left to right through the external circuit of the cell, as indicated in Figure 22. If this takes place *under the reversible conditions outlined above*, then the potential difference between the electrodes is the cell emf, and the difference in electrical energy is just the electrical work done. Under these circumstances, equation 74 becomes

$$w_{\mathrm{el,\,max}} = E\mathrm{d}Q \qquad (75)$$

where (as before) $w_{\mathrm{el,\,max}}$ is the work done *on* the system. Now suppose that the charge that flows consists of $\mathrm{d}n$ moles of electrons. The charge carried by one mole of *protons* (positive particles) is just the **Faraday constant** F (= 96 485 C mol^{-1}); that carried by $\mathrm{d}n$ moles of *electrons* (negative particles) must then be

$$\mathrm{d}Q = -F\mathrm{d}n$$

so that equation 75 becomes

$$w_{\mathrm{el,\,max}} = -FE\mathrm{d}n \qquad (76)$$

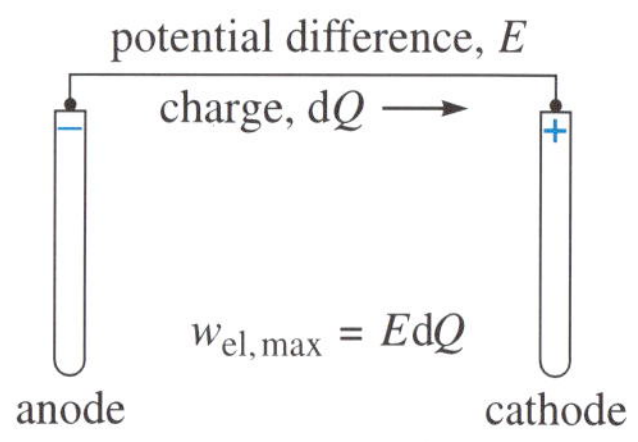

Figure 22 The transfer of charge $\mathrm{d}Q$ through a potential difference E, under reversible conditions.

■ Does equation 76 accord with the sign convention for E introduced in Section 2.3.2?

■ Yes. Work can be *obtained* from the cell (that is, $w_{el, max}$ is negative) only if electrons flow *spontaneously* from left to right in the external circuit of Figure 22. This requires a positive emf ($E > 0$) in equation 76, in line with the sign convention for E.

The final step is to relate this transfer of dn moles of electrons to the underlying cell reaction:

$$Cu(s) + 2Ag^+(aq) = Cu^{2+}(aq) + 2Ag(s) \qquad (8)$$

■ According to the stoichiometry of equation 8, how many moles of electrons are produced for each mole of copper oxidized at the anode?

■ Two, as implied by the half-reaction at the anode:

$$Cu(s) = Cu^{2+}(aq) + 2e \qquad (9)$$

■ If you now think of 2 as the 'stoichiometric number' of the electron in equation 9, write an expression for dn in terms of the corresponding change dξ in the extent of this reaction. (Refer back to the discussion in Section 4.3 if necessary.)

■ In this case, $\mathrm{d}n = 2\mathrm{d}\xi$.

In general, the coefficient of e (implicit in the cell reaction as written) is given the symbol n.* Then equation 76 becomes,

$$w_{el, max} = -nFE\mathrm{d}\xi \qquad (77)$$

which can be combined with equation 73 and reorganized to give:

$$(\mathrm{d}G/\mathrm{d}\xi) = -nFE \qquad (78)$$

This, then, is the formal connection between $(\mathrm{d}G/\mathrm{d}\xi)$ and E that we set out to obtain.

■ Check again that equation 78 accords with the sign convention for E introduced in Section 2.3.2.

■ Yes it does. For example, the reaction in equation 8 will proceed from left to right if $(\mathrm{d}G/\mathrm{d}\xi) < 0$. According to the sign convention in Section 2.3.2, this corresponds to a *positive* emf for the cell in Figure 21, in line with equation 78.

SAQ 14 Consider again the cell you examined in SAQ 8:

$$Pt, H_2(g)|H^+(aq)|Ag^+(aq)|Ag(s)$$

with the implied cell reaction:

$$\tfrac{1}{2}H_2(g) + Ag^+(aq) = H^+(aq) + Ag(s) \qquad (16)$$

The values of $(\mathrm{d}G/\mathrm{d}\xi)$ you calculated in part (b) of SAQ 8 are included in the table below. Use equation 78 to calculate the corresponding values of E, and hence confirm the statement made above. What approximation is inherent in these calculations?

$c(H^+)$/mol dm^{-3}	$c(Ag^+)$/mol dm^{-3}	$(\mathrm{d}G/\mathrm{d}\xi)$/kJ mol^{-1}
(i) 10^{-5}	10^{-1}	−99.9
(ii) 10^{-5}	10^{-21}	+14.2

* It is unfortunate (to say the least!) that the *same* symbol is conventionally used to represent both the *amount of substance* (in moles) – the amount dn of electrons transferred, for instance – and the 'stoichiometric coefficient' of the electron, which is just a number. Try not to confuse the two uses.

5.3 The Nernst equation

Of the questions posed in Section 2, only one remains unanswered – namely the concentration- (or more accurately, activity-) dependence of the cell emf. Consider again the Daniell cell:

$$Zn(s)|Zn^{2+}(aq)|Cu^{2+}(aq)|Cu(s)$$

with

$$Zn(s) + Cu^{2+}(aq) = Zn^{2+}(aq) + Cu(s) \qquad (67)$$

As you should have found in answering Exercise 2, when the cell is 'under standard conditions' (that is, each of the active cell components is at unit activity), then the final term in equation 51 becomes zero:

$$(dG/d\xi) = \Delta G_m^\ominus \quad \text{(under standard conditions)} \qquad (79)$$

The corresponding emf is called the **standard emf**, and denoted $E^\ominus$: it is related to the value of $\Delta G_m^\ominus$ for the implied cell reaction (equation 67, in this case), as follows:

$$\Delta G_m^\ominus = -nFE^\ominus \qquad (80)$$

- At 298.15 K, $\Delta G_m^\ominus = -212.6\,\text{kJ mol}^{-1}$ for the reaction in equation 67. What is the corresponding value of $E^\ominus$?

- $n = 2$ (from equation 67), so

$$E^\ominus = \frac{-\Delta G_m^\ominus}{nF} = \frac{(-212.6 \times 10^3\ \text{J mol}^{-1})}{(2 \times 96\ 485\ \text{C mol}^{-1})} = 1.10\,\text{J C}^{-1} = 1.10\,\text{V}$$

Alternatively, you can think of equation 80 as the *definition* of $E^\ominus$. Either way, equations 78 and 80 can be substituted back into equation 51 to give

$$-nFE = -nFE^\ominus + RT\ln Q$$

which is usually reorganized as follows:

$$E = E^\ominus - (RT/nF)\ln Q \qquad \textbf{Nernst equation} \qquad (81)$$

Equation 81 is named after the man whose brilliant analysis first led to its derivation (Figure 23). Evidently, it relates the emf E under arbitrary conditions to the standard value $E^\ominus$: *it is one of the key results in this Block*, and we shall put it to good use in subsequent Sections.

To summarize: culminating in the Nernst equation, the discussion in this Section has provided a formal basis for the central theme running through this Block. For many chemical systems – and in particular, those involving aqueous ions – the emf is an equally good measure of the 'tendency for reaction'. As far as thermodynamics is concerned, whether such a system happens to be set up in an electrochemical cell, or a test-tube, or wherever, is completely irrelevant. In particular, the standard emf, $E^\ominus$, is really just an alternative way of expressing the value of $\Delta G_m^\ominus$ for a particular *reaction*. From this perspective, the prime importance of electrochemical cells lies in the fact that they can provide an alternative, *non-calorimetric*, source of this and other thermodynamic data. These points are taken up in the next two Sections.

On the other hand, electrochemical cells obviously do have immense practical importance in their own right – not only as valuable sources of electricity (batteries, for instance), but also in providing a means of preparing substances (via electrolysis) that would be difficult, if not impossible, to obtain in other ways. These more practical aspects of electrochemistry are discussed further in Section 8.

Figure 23 Walther Nernst (1864–1941), received the Nobel Prize for Chemistry in 1920 for his thermochemical work. Among many other interests, he was an enthusiastic carp farmer, arguing on the basis of second-law considerations that fish were a better investment from an energy standpoint than warm-blooded livestock. (Courtesy Carlsberg Foundation)

6 ELECTROCHEMICAL CELLS AND SOLUTION REACTIONS

We are now in a position to draw together the threads running through previous Sections. First of all, we introduce you to the way standard emfs are recorded in the chemical literature. We then go on to examine the use of those data in a variety of different contexts – only some of which have to do with electrochemical cells, as such. For this reason, a more detailed discussion of the measurement of cell emfs – by direct electrochemical means – is deferred until Section 7.

6.1 Standard electrode potentials

As with other types of thermodynamic data, the task of trying to list the standard emf for every cell (essentially, every reaction) that has been studied is an awesome prospect. Fortunately, there is a way around this problem – one that makes it possible to have just one entry for each *half-reaction*, its *standard electrode potential.*

6.1.1 Redox couples: defining the standard electrode potential

To begin with, consider again the cell discussed in Section 3 (Figure 10), as represented by the following cell diagram:

$$\mathrm{Pt,\ H_2(g)|H^+(aq)|Ag^+(aq)|Ag(s)}$$

At that stage, the implied cell reaction was arrived at by applying the 'rule' introduced in Section 2.3.1 (oxidation at the LHE combined with reduction at the RHE), that is:

RHE (reduction):	$\mathrm{Ag^+(aq) + e = Ag(s)}$	(10)
LHE (oxidation):	$\mathrm{\tfrac{1}{2}H_2(g) = H^+(aq) + e}$	
Add	$\mathrm{\tfrac{1}{2}H_2(g) + Ag^+(aq) = H^+(aq) + Ag(s)}$	(16)

For our present purposes, however, it is crucial to appreciate that equation 16 can equally well be thought of as the *difference between two reduction processes*, as follows:

RHE (reduction):	$\mathrm{Ag^+(aq) + e = Ag(s)}$	(10)
LHE (reduction):	$\mathrm{H^+(aq) + e = \tfrac{1}{2}H_2(g)}$	
Subtract (RHE – LHE):	$\mathrm{Ag^+(aq) - H^+(aq) = Ag(s) - \tfrac{1}{2}H_2(g)}$	

which can be reorganized to give equation 16.

This alternative interpretation of the implied cell reaction is the starting point for the definition of the standard electrode potential of the half-reaction, or *couple*, in equation 10 – which can be represented more concisely as ($\mathrm{Ag^+(aq)|Ag(s)}$), or just ($\mathrm{Ag^+|Ag}$).

The first step is to suppose that the cell is set up 'under standard conditions' – that is, *each of the active components is taken to be at unit activity* – such that the emf of the cell has its standard value, $E^{\ominus}$(cell):*

$$\mathrm{Pt,\ H_2(g)\ |\ H^+(aq)\ |\ Ag^+(aq)\ |\ Ag(s)}$$

$$p = 1\ \mathrm{bar}\quad a(\mathrm{H^+}) = 1\quad a(\mathrm{Ag^+}) = 1\quad a(\mathrm{Ag}) = 1$$

* As we hinted in the answer to Exercise 2, the best we could do *in practice* would be to arrange matters such that the *mean* ionic activities of the electrolytes in the two half-cells are both unity. Strictly speaking, then, this cell *cannot* actually be set up 'under standard conditions'. Nevertheless, its standard emf *can* be determined experimentally – using the sort of analysis outlined in Section 7.2. For now, we ignore this complication.

Under these circumstances, the half-cell on the left-hand side is called the **standard hydrogen electrode (S.H.E.)**: effectively, it comprises a piece of platinum metal bathed in hydrogen gas at a pressure of 1 bar, and immersed in an aqueous solution containing H^+(aq) ions at unit activity (Figure 24).

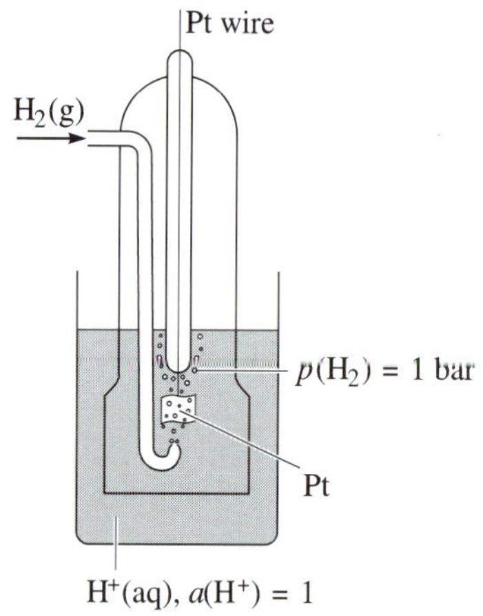

Figure 24 The standard hydrogen electrode.

Next, the standard emf of the cell is divided into two 'electrode potentials', as follows:

$$E^\ominus(\text{cell}) = E^\ominus_{\text{RHE}} - E^\ominus_{\text{LHE}} \qquad (82)$$

where $E^\ominus_{\text{RHE}}$ and $E^\ominus_{\text{LHE}}$ now refer to the half-reactions at the right-hand and left-hand electrode, respectively, *each written as a reduction*, as:

$$E^\ominus_{\text{RHE}} = E^\ominus(\text{Ag}^+|\text{Ag}) \text{ and } E^\ominus_{\text{LHE}} = E^\ominus(\text{H}^+|\text{H}_2)$$

where the shorthand $(H^+|H_2)$ represents the reduction process $H^+(aq) + e = \frac{1}{2}H_2(g)$.

In the final step, *the electrode potential of the standard hydrogen electrode is arbitrarily assigned the value zero*, that is:

S.H.E. $\quad E^\ominus(\text{H}^+|\text{H}_2) = 0$ (by definition) $\qquad$ (83)

The force of this *arbitrary convention* is to assign the whole of the standard emf of the cell to the $(Ag^+|Ag)$ couple – that is, equation 82 becomes

$$E^\ominus(\text{cell}) = E^\ominus(\text{Ag}^+|\text{Ag})$$

Or to put it more generally:

The **standard electrode potential** of a half-cell (or half-reaction, or couple) is *defined* as the standard emf of a cell in which the S.H.E. is on the left and the half-cell under investigation is on the right.

■ There was a reminder in Section 3 that equation 16 is the *reverse* of the formation reaction for Ag^+(aq), such that $\Delta G^\ominus_{\text{m}}(16) = -\Delta G^\ominus_{\text{f}}(\text{Ag}^+, \text{aq}) = -77.1\ \text{kJ mol}^{-1}$ at 298.15 K. Use the relation $\Delta G^\ominus_{\text{m}} = -nFE^\ominus$ (equation 80 in Section 5.3) to calculate $E^\ominus(\text{Ag}^+|\text{Ag})$ at 298.15 K.

■ Using equation 80, the standard emf of the reaction in equation 16 is given by $E^\ominus = -\Delta G^\ominus_{\text{m}}(16)/nF$, with $n = 1$, as

$$E^\ominus = \frac{-(-77.1\times10^3\ \text{J mol}^{-1})}{(1\times96\,485\ \text{C mol}^{-1})} = +0.80\ \text{J C}^{-1}$$

$$= +0.80\ \text{V}$$

But this is the standard emf of the cell discussed above. Hence it is also the value of $E^\ominus(\text{Ag}^+|\text{Ag})$.

■ The assignment $E^{\ominus} = 0$ to the S.H.E. is equivalent to another arbitrary convention that you met in the Second Level Inorganic Course. Can you spot the connection?

■ Assigning zero to the value $E^{\ominus}(H^+|H_2)$ allows $E^{\ominus}$ for the *reaction* in equation 16 to be ascribed to the couple $(Ag^+|Ag)$. Calculating the corresponding value of $\Delta G_m^{\ominus}$ involves the equally arbitrary, but internally consistent, assignment:

$$\Delta G_f^{\ominus}(H^+, aq) = 0 \text{ (by definition)}$$

STUDY COMMENT Make sure you try the following SAQ before moving on.

SAQ 15 (a) Use information from Section 2 of the S342 *Data Book* to calculate the standard electrode potential at 298.15 K of the following couple:

$$Zn^{2+}(aq) + 2e = Zn(s)$$

In view of the discussion above, what is the significance of the *sign* of $E^{\ominus}$ for this couple?

(b) Would your calculated value of $E^{\ominus}$ be affected by writing the couple with the following stoichiometry?

$$\tfrac{1}{2}Zn^{2+}(aq) + e = \tfrac{1}{2}Zn(s)$$

6.1.2 Standard electrode potentials as part of the thermodynamic database

Once standard electrode potentials have been determined (either by experiment – more on which in Section 7 – or more often, by calculation from other thermodynamic data), they are tabulated in the chemical literature, generally for results at a temperature of 298.15 K (25 °C). Table 5 lists a selection of such values for couples involving metals and their aqueous ions: notice that it includes the values for $(Ag^+|Ag)$ and $(Zn^{2+}|Zn)$ calculated above. You will find a more comprehensive compilation in the S342 *Data Book*.

Table 5 Standard electrode potentials at 298.15 K.[a]

Electrode reaction	$E^{\ominus}$/V
$K^+(aq) + e = K(s)$	−2.94
$Ca^{2+}(aq) + 2e = Ca(s)$	−2.87
$Na^+(aq) + e = Na(s)$	−2.71
$Mg^{2+}(aq) + 2e = Mg(s)$	−2.36
$Al^{3+}(aq) + 3e = Al(s)$	−1.68
$Zn^{2+}(aq) + 2e = Zn(s)$	−0.76
$Cr^{3+}(aq) + 3e = Cr(s)$	−0.74
$Fe^{2+}(aq) + 2e = Fe(s)$	−0.46
$Cd^{2+}(aq) + 2e = Cd(s)$	−0.40
$Ni^{2+}(aq) + 2e = Ni(s)$	−0.24
$Sn^{2+}(aq) + 2e = Sn(s)$	−0.14
$Pb^{2+}(aq) + 2e = Pb(s)$	−0.13
$H^+(aq) + e = \tfrac{1}{2}H_2(g)$	0
$Cu^{2+}(aq) + 2e = Cu(s)$	+0.34
$Ag^+(aq) + e = Ag(s)$	+0.80
$Hg_2^{2+}(aq) + 2e = 2Hg(l)$	+0.80
$Hg^{2+}(aq) + 2e = Hg(l)$	+0.85

[a] Values have mostly been calculated from the thermodynamic data in Section 2 of the S342 *Data Book*.

There are several points to note about the data in this Table. First, in line with the definition in the previous Section, each of the entries is written as a reduction – with the oxidized state (e.g. Mg^{2+}) on the left, and the reduced state (e.g. Mg) on the right. Thus, a concise way of expressing each couple is to write $E^{\ominus}(M^{n+}|M)$, with the oxidized state first inside the bracket, and the reduced state second – in line with the rule for writing the half-cell on the right-hand side of a cell diagram introduced in Section 2.3.1. Values of $E^{\ominus}$ defined in this way are sometimes referred to as **standard reduction potentials***. Two further points were highlighted by the example in SAQ 15.

Specifically, part (a) serves to emphasize the role of the arbitrary convention introduced above. Strictly speaking, the symbol 'e' in Table 5 is simply shorthand for $\{\tfrac{1}{2}H_2(g) - H^+(aq)\}$ or 'add $\tfrac{1}{2}H_2(g)$ to the left-hand side and $H^+(aq)$ to the right'. It follows that in this context e is *not* a symbol for the electron, although it does balance the equations with respect to charge if one regards it as such.

What, then, is the significance of the sign of $E^{\ominus}$ for the couples in Table 5?

* This convention accords with the recommendations of the International Union of Pure and Applied Chemistry (IUPAC). If you consult older texts (especially ones published in the United States), you may encounter tables in which all the signs are opposite to those given here. This reflects the adoption of a different convention, in which *standard oxidation potentials* are listed. It is vital to check which convention is being used.

Under standard conditions, $E^{\ominus}$ provides a measure of the *thermodynamic stability* of a metal with respect to oxidation by $H^+(aq)$ to the aqueous cation. Thus, any metal more readily oxidized than $H_2(g)$ will have a negative value for $E^{\ominus}$, and vice versa.

The reason why $E^{\ominus}$ is such a convenient measure of the oxidizing or reducing power of a particular couple revolves around the point raised in SAQ 15b:

> The value of $E^{\ominus}$ is independent of the way in which we choose to write the stoichiometry of the half-reaction.

This is important. Its import can be underlined by considering the following reaction:

$$Zn(s) + 2Ag^+(aq) = Zn^{2+}(aq) + 2Ag(s) \quad (84)$$

Thinking of this as an implied cell reaction, it can be 'decomposed' into the difference between two couples 'RHE – LHE', with:

RHE: $2Ag^+(aq) + 2e = 2Ag(s)$

LHE: $Zn^{2+}(aq) + 2e = Zn(s)$

- Using information from Table 5, is reaction 84 thermodynamically favourable under standard conditions at 298.15 K?

- Yes, since $E^{\ominus} > 0$. The crucial point is that here $E^{\ominus}_{RHE} = E^{\ominus}(Ag^+|Ag)$, as listed in Table 5, even though the stoichiometry of the half-reaction above is *double* that written in Section 2.1 (equation 10). Thus:

$$\begin{aligned} E^{\ominus} &= E^{\ominus}_{RHE} - E^{\ominus}_{LHE} \\ &= E^{\ominus}(Ag^+|Ag) - E^{\ominus}(Zn^{2+}|Zn) \\ &= \{+0.80 - (-0.76)\}\text{ V} = 1.56\text{ V} \end{aligned}$$

The conclusion that reaction 84 does represent a thermodynamically favourable process could be arrived at with the minimum of fuss by simply noting that the equations in Table 5 are arranged in order of increasing values of $E^{\ominus}$, with the most negative values at the top, the most positive at the bottom. Now a reaction (equation 84, say) will go from left to right as written, with reduction of the 'right-hand couple' ($Ag^+|Ag$, in this case) if $E^{\ominus} > 0$. And $E^{\ominus}$ is positive when $E^{\ominus}_{RHE} - E^{\ominus}_{LHE} > 0$ – that is, when $E^{\ominus}_{RHE}$ is *more positive* than $E^{\ominus}_{LHE}$ (as in the example above). It follows that metal ions lower down Table 5 are more readily reduced than those higher up. Or to put it another way, any metal is thermodynamically capable (under standard conditions at 298.15 K) of reducing the ions of another metal below it to that metal itself, the first metal being simultaneously oxidized to its ions.

Table 5 is, then, an equally good expression of the thermodynamic basis for the **electrochemical series** of metals that you met in the Second Level Inorganic Course. The ramifications of this approach to the 'reactivity' of metals were examined in detail at that stage: we shall not consider the problem further in this Course – other, that is, than to note again the limitations inherent in *any* treatment based solely on thermodynamic arguments (see the following SAQ).

SAQ 16 Use Table 5 to predict which of the following reactions are thermodynamically favourable under standard conditions at 298.15 K: (a) reduction of $Zn^{2+}(aq)$ by magnesium; (b) reduction of $H^+(aq)$ by silver; (c) reduction of $Cu^{2+}(aq)$ by aluminium. Are your predictions consistent with the experimental observation that the reduction in (a) is the only one that actually occurs under ambient conditions?

6.1.3 Extending the database

Even leaving aside the possibility of unknown kinetic effects, as we shall for the present, the discussion in the previous Section is still restricted in one important respect: couples need not contain just metals and their aqueous cations. Indeed, as a quick glance through your S342 *Data Book* will confirm, almost any oxidized and reduced state can form a couple: a selection of representative examples is collected in Table 6. Many of these systems *cannot* be studied directly in electrochemical cells: their electrode potentials must be obtained by indirect methods. Exceptions include the simpler couples containing two different oxidation states of a metal, for instance,

$$Fe^{3+}(aq) + e = Fe^{2+}(aq) \quad (85)$$

and the so-called **metal/insoluble salt electrodes**, for example the silver/silver chloride electrode:

$$AgCl(s) + e = Ag(s) + Cl^{-}(aq) \quad (86)$$

Suitable half-cells are indicated in Figures 25 and 26, respectively.

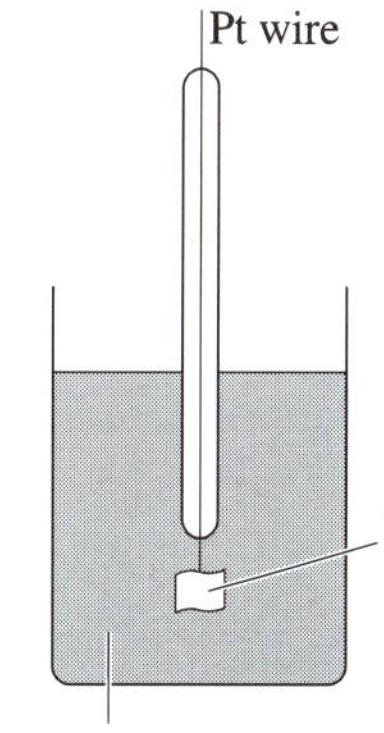

Figure 25 The ($Fe^{3+}|Fe^{2+}$) redox couple. It comprises an inert (usually platinum) electrode in contact with a solution containing iron in two oxidation states, $Fe^{2+}(aq)$ and $Fe^{3+}(aq)$.

Table 6 Standard electrode potentials at 298.15 K.[a]

Electrode reaction	$E^{\ominus}/V$
$Ag_2S(s) + 2e = 2Ag(s) + S^{2-}(aq)$	−0.66
$PbSO_4(s) + 2e = Pb(s) + SO_4^{2-}(aq)$	−0.36
$H^+(aq) + e = \frac{1}{2}H_2(g)$	0
$AgBr(s) + e = Ag(s) + Br^-(aq)$	0.07
$Sn^{4+}(aq) + 2e = Sn^{2+}(aq)$	0.15
$AgCl(s) + e = Ag(s) + Cl^-(aq)$	0.22
$\frac{1}{2}I_2(s) + e = I^-(aq)$	0.53
$Fe^{3+}(aq) + e = Fe^{2+}(aq)$	0.77
$\frac{1}{2}Br_2(l) + e = Br^-(aq)$	1.08
$\frac{1}{2}O_2(g) + 2H^+(aq) + 2e = H_2O(l)$	1.23
$Cr_2O_7^{2-}(aq) + 14H^+(aq) + 6e = 2Cr^{3+}(aq) + 7H_2O(l)$	1.36
$\frac{1}{2}Cl_2(g) + e = Cl^-(aq)$	1.36
$MnO_4^-(aq) + 8H^+(aq) + 5e = Mn^{2+}(aq) + 4H_2O(l)$	1.51
$PbO_2(s) + SO_4^{2-}(aq) + 4H^+(aq) + 2e = PbSO_4(s) + 2H_2O(l)$	1.69
$\frac{1}{2}F_2(g) + e = F^-(aq)$	2.89

[a] As in Table 5, values have mostly been calculated from other thermodynamic data.

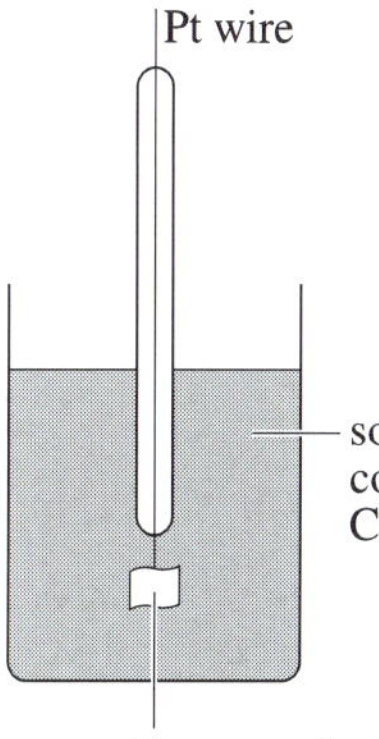

Figure 26 The silver/silver chloride electrode. It comprises metallic silver covered by a layer of the 'insoluble' salt, AgCl, the whole immersed in a solution containing the anion, $Cl^-(aq)$. Systems like this have an important role in experimental electrochemistry – more on which in Section 7.

As the discussion in the previous Section suggests, the impetus for such extensive compilations of electrode potential data stems, in part at least, from the ease with which they can be combined in order to predict the thermodynamic favourability of a given reaction. In which context – and to generalize our earlier discussion:

> Given two redox couples, where the $E^{\ominus}$ value for the first is more positive than that for the second, then the oxidized state in the first is thermodynamically capable of oxidizing the reduced state in the second.

Obviously, the inclusion of more 'exotic' couples, like the ones in Table 6, increases enormously the number, and type, of reactions that can be handled in this way. More often than not, the problem now reduces to identifying the two couples that will, when combined, yield the desired process. *If you would like some practice at this stage, try the following SAQ.*

SAQ 17 Use information from Table 6 to predict which of the following reactions are thermodynamically favourable under standard conditions at 298.15 K:

(a) $H^+(aq) + Cl^-(aq) = \frac{1}{2}H_2(g) + \frac{1}{2}Cl_2(g)$

(b) $2Fe^{3+}(aq) + Sn^{2+}(aq) = 2Fe^{2+}(aq) + Sn^{4+}(aq)$

(c) $6Fe^{2+}(aq) + Cr_2O_7^{2-}(aq) + 14H^+(aq) = 6Fe^{3+}(aq) + 2Cr^{3+}(aq) + 7H_2O(l)$

(d) $PbO_2(s) + Pb(s) + 4H^+(aq) + 2SO_4^{2-}(aq) = 2PbSO_4(s) + 2H_2O(l)$

6.2 Using electrode potential data: a broader perspective

So far, we have concentrated on the definition and use of standard electrode potentials – a discussion that has paid scant attention to a fundamental restriction on the use of such data: they relate to *standard* conditions. If the chemical system is actually set up in an electrochemical cell, this implies that each of the ionic species, whether 'reactant' or 'product', is present at unit activity. Achieving this *in practice* raises problems of its own – an issue that is taken up in Section 7. Nevertheless, it should be clear that tabulated $E^{\ominus}$ values can be used to calculate the standard emf of such a cell, and hence to predict the spontaneous cell reaction under standard conditions. (See the SAQs in Section 6.3 for examples of this type of calculation.) But how do predictions based on $E^{\ominus}$ data relate to the normal laboratory situation, where reactants are simply mixed together and the reaction is allowed to run its course? This is the question we turn to now.

6.2.1 Electrode potentials and equilibrium constants

The question posed above is one that was touched on in Exercise 2. In the present context, it is best examined by reference to the Nernst equation, which relates the emf E under arbitrary conditions to the standard value $E^{\ominus}$:

$$E = E^{\ominus} - (RT/nF)\ln Q \qquad (81)$$

- According to equation 81, what is the relation between $E^{\ominus}$ and the standard equilibrium constant $K^{\ominus}$ for a particular reaction?

- Like its counterpart $(\mathrm{d}G/\mathrm{d}\xi)$, E is zero at equilibrium: the ratio of activities in the reaction quotient Q then takes its equilibrium value, and equation 81 becomes:

  $$0 = E^{\ominus} - (RT/nF)\ln K^{\ominus}$$

or

$$E^{\ominus} = (RT/nF)\ln K^{\ominus} \qquad (87)$$

Thus $E^{\ominus}$, like $\Delta G_{\mathrm{m}}^{\ominus}$, is really just a measure of the standard equilibrium constant for a reaction. As an example, consider the reaction in equation 88:

$$Sn^{2+}(aq) + 2Fe^{3+}(aq) = Sn^{4+}(aq) + 2Fe^{2+}(aq) \qquad (88)$$

- Given that $E^{\ominus} = 0.62$ V at 298.15 K, write an expression for $K^{\ominus}$ for this reaction, and calculate its value.

- $\ln K^{\ominus} = (nFE^{\ominus}/RT)$, with $n = 2$ (see SAQ 17b)

  $= (2 \times 96\,485\ \mathrm{C\,mol^{-1}} \times 0.62\ \mathrm{V})/(8.314\ \mathrm{J\,K^{-1}\,mol^{-1}} \times 298.15\ \mathrm{K})$

  $= 48.266$

  So $K^{\ominus} = \dfrac{\{a(Sn^{4+})\}_e\{a(Fe^{2+})\}_e^2}{\{a(Sn^{2+})\}_e\{a(Fe^{3+})\}_e^2} = 9.15 \times 10^{20}$

This large value of $K^{\ominus}$ suggests that equilibrium lies well over to the right-hand side in this system – but there is an important proviso. The *standard* equilibrium constant is a function of activities, but it is the equilibrium position *in terms of concentrations* that interests the practising chemist.

- Using the symbol K_c to denote the 'dimensionless' equilibrium constant in terms of concentrations, write an expression relating $K^{\ominus}$ and K_c for reaction 88. (It may help to refer back to the discussion in Section 4.7.1.)

- Using the definition of activity for a solute in a real solution (equation 55), the expression for $K^{\ominus}$ becomes:

$$K^{\ominus} = \frac{\{\gamma(\mathrm{Sn}^{4+})c(\mathrm{Sn}^{4+})/c^{\ominus}\}_e \{\gamma(\mathrm{Fe}^{2+})c(\mathrm{Fe}^{2+})/c^{\ominus}\}_e^2}{\{\gamma(\mathrm{Sn}^{2+})c(\mathrm{Sn}^{2+})/c^{\ominus}\}_e \{\gamma(\mathrm{Fe}^{3+})c(\mathrm{Fe}^{3+})/c^{\ominus}\}_e^2}$$

$$= \frac{\{c(\mathrm{Sn}^{4+})/c^{\ominus}\}_e \{c(\mathrm{Fe}^{2+})/c^{\ominus}\}_e^2}{\{c(\mathrm{Sn}^{2+})/c^{\ominus}\}_e \{c(\mathrm{Fe}^{3+})/c^{\ominus}\}_e^2} \times \left(\frac{\gamma(\mathrm{Sn}^{4+})\{\gamma(\mathrm{Fe}^{2+})\}^2}{\gamma(\mathrm{Sn}^{2+})\{\gamma(\mathrm{Fe}^{3+})\}^2}\right)$$

Or more simply,

$K^{\ominus} = K_c \times$ (function of activity coefficients)

As you saw in Section 4.7.4, strictly speaking it is only at 'infinite dilution' (that is, at very low ionic strengths) that activity coefficients become unity, so that the concentration and standard equilibrium constants coincide. Happily, it turns out that the activity coefficients of common electrolytes rarely differ by more than a factor of ten in normal aqueous media. Thus, $K^{\ominus}$ (and hence $E^{\ominus}$) is usually a good guide (within a factor of 100 or so) to the value of K_c for a reaction – and thence to the observed equilibrium position.

With this in mind, concentrate on the information in Table 7: it contains a range of $E^{\ominus}$ values together with the corresponding (according to equation 87) values of $K^{\ominus}$ at 298.15 K. All of the values in the Table relate to reactions involving transfer of only one electron (that is, $n = 1$ in equation 87). Nevertheless, it is clear that many redox reactions can be expected to have either very small or very large values for the equilibrium constant. For example, a difference in $E^{\ominus}$ for two couples of just +0.5 V is more than sufficient to ensure that the combined reaction is effectively complete at 298.15 K.

Table 7 Some numerical values of $E^{\ominus}$ together with corresponding values of $K^{\ominus}$ at 298.15 K, calculated from equation 87, with $n = 1$.

$E^{\ominus}$/V	$K^{\ominus}$
1.0	8.0×10^{16}
0.5	2.8×10^{8}
0.1	4.9×10^{1}
0	1.0
–0.1	2.0×10^{-2}
–0.5	3.5×10^{-9}
–1.0	1.2×10^{-17}

How do these comments relate to the problem of predicting the direction of spontaneous change in a chemical system of arbitrary composition?

They suggest strongly that, for many redox systems, *the size (and sign!) of $E^{\ominus}$ is often sufficient to dominate the expression on the right-hand side of equation 81. Under these circumstances, the sign of $E^{\ominus}$ may be taken as a sufficient criterion of spontaneous change*. You may care to check the answers to SAQs 16 and 17 in the light of this discussion. (See also the answer to Exercise 2.)

As a postscript, equation 87 also highlights another important application of standard electrode potentials, namely the determination of standard equilibrium constants, many of which are difficult, if not impossible, to measure directly. Indeed, Tables 5 and 6 contain in a concise form all the material necessary for the calculation of 435 such equilibrium constants.

SAQ 18 Use information from Tables 5 and 6 to determine the standard solubility product, $K_{sp}^{\ominus}$, for AgCl at 298.15 K. *Hint* Look for two couples that will, when combined, yield the following overall reaction:

$$AgCl(s) = Ag^{+}(aq) + Cl^{-}(aq) \quad (52)$$

6.2.2 Cell emf and the Nernst equation

Despite the discussion in the previous Section, there are situations that do require use of the full Nernst equation. One obvious example is where interest centres around the actual emf delivered by a particular cell for some stated, but non-standard, activities of its active components. Indeed, it was a system like this that led us to embark on the thermodynamic analysis in Sections 4 and 5.

Here, we consider briefly a more practical example: the lead/acid battery (or lead accumulator) that is used to start a car. Basically, it comprises an anode of metallic lead and a cathode of lead(IV) oxide, PbO_2, together with an electrolyte of dilute sulfuric acid (Figure 27).

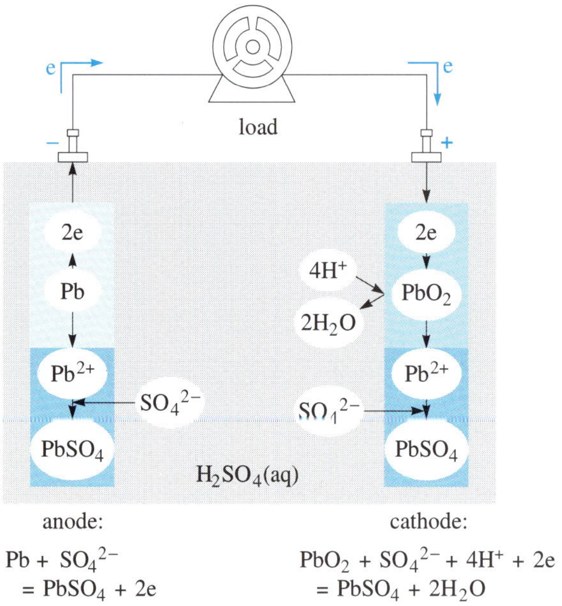

Figure 27 Schematic representation of the spontaneous (discharge) reaction in a cell of the lead/acid battery. (Notice that the solid discharge product at *both* electrodes is lead sulfate.)

The overall cell reaction can be represented as follows:

$$Pb(s) + PbO_2(s) + 4H^+(aq) + 2SO_4^{2-}(aq) = 2PbSO_4(s) + 2H_2O(l) \qquad (89)$$

with $E^{\ominus} = +2.05$ V at 298.15 K (as calculated in SAQ 17d).

Suppose now that the acid (H_2SO_4) is present at a *concentration* of 0.5 mol dm^{-3}: what is the corresponding emf of the cell? The first step is to use the Nernst equation (equation 81) to write an expression for E.

■ Write this down now, remembering that $n = 2$, in equation 89 (Figure 27).

$$E = 2.05\text{ V} - \frac{RT}{2F}\ln\left(\frac{\{a(PbSO_4)\}^2\{a(H_2O)\}^2}{a(Pb)\,a(PbO_2)\{a(H^+)\}^4\{a(SO_4^{2-})\}^2}\right)$$

This expression can be simplified by recalling that the activities of the solids and the solvent (H_2O) can be taken as unity, whence

$$E = 2.05\text{ V} - \frac{RT}{2F}\ln\{a(H^+)\}^{-4}\{a(SO_4^{2-})\}^{-2}$$

$$= 2.05\text{ V} + \frac{RT}{F}\ln\{a(H^+)\}^2 a(SO_4^{2-}) \qquad (90)$$

STUDY COMMENT If you are not overly confident about handling logs, then it would be a good plan to proceed in steps. Here, for example, the logarithmic term can be split into two, as

$$-\frac{RT}{2F}(\ln\{a(H^+)\}^{-4} + \ln\{a(SO_4^{2-})\}^{-2})$$

$$= -\frac{RT}{2F}(-4\ln\{a(H^+)\} - 2\ln\{a(SO_4^{2-})\}), \quad \text{since } \ln x^{-a} = -a\ln x$$

$$= +\frac{RT}{F}(2\ln\{a(H^+)\} + \ln\{a(SO_4^{2-})\}) = +\frac{RT}{F}(\ln\{a(H^+)\}^2 + \ln\{a(SO_4^{2-})\})$$

When these two logarithmic terms are recombined they give the term in equation 90.

The final step is to relate the activities in equation 90 to the stated concentration of the acid. In this case, the requisite information, namely the mean ionic activity coefficient $\gamma_{\pm}$ for 0.5 molar sulfuric acid, has been measured experimentally. The value is given in Table 4 as $\gamma_{\pm} = 0.155$, and it relates activities to concentrations via an expression that is a generalization of the 1 : 1 case you met in Section 4.7.2 (see Table 3):

$$\{a(H^+)\}^2 a(SO_4^{2-}) = \{\gamma(H^+)c(H^+)/c^{\ominus}\}^2\{\gamma(SO_4^{2-})c(SO_4^{2-})/c^{\ominus}\}$$

$$= \gamma_{\pm}^3\{c(H^+)/c^{\ominus}\}^2\{c(SO_4^{2-})/c^{\ominus}\}$$

In this instance, $c(SO_4^{2-}) = 0.5\ mol\ dm^{-3}$; $c(H^+) = 2c(SO_4^{2-}) = 1.0\ mol\ dm^{-3}$ and $\gamma_\pm = 0.155$, so equation 90 becomes:

$$E = 2.05\ V + \left(\frac{8.314 \times 298.15}{96\,485}\right) V \ln (0.155)^3(1.0)^2(0.5)$$

$$= (2.05 - 0.16)\ V$$

$$= 1.89\ V$$

which is significantly different from the standard value.

This example was chosen with care. Not only does it demonstrate the use of the Nernst equation in a practical context, but it also serves to underline the, sometimes gross, approximations inherent in identifying activities with molar concentrations. Thus, in this case, the predicted emf would be 2.03 V without the 'correction factor' contained in $\gamma_\pm$! Nevertheless, for want of the necessary data, we shall often be obliged to make this approximation (see the SAQs in Section 6.3 for examples).

6.2.3 Electrode potentials and the Nernst equation

With the proviso that the values of E thereby obtained are still referred to the standard hydrogen electrode, the Nernst equation can equally well be applied to *individual* couples. Here we look briefly at two situations in which the shift to 'non-standard' conditions can have a marked effect on the behaviour of a particular couple.

Couples that involve the hydrogen (or hydroxide) ion

For such couples, the electrode potential depends on the concentration (strictly, activity) of hydrogen ions – which may vary by a factor of about 10^{14} in aqueous solution. As an example, consider the following couple (taken from Table 6):

$$\tfrac{1}{2}O_2(g) + 2H^+(aq) + 2e = H_2O(l);\quad E^\ominus = 1.23\ V \tag{91}$$

- Use the Nernst equation to write an expression for the electrode potential of this couple.

$$E = E^\ominus - \frac{RT}{2F}\ln\left(\frac{a(H_2O)}{\{a(O_2)\}^{1/2}\{a(H^+)\}^2}\right) \tag{92}$$

If we write $a(H_2O) = 1$, and assume, as we shall throughout this Section, that the pressure of oxygen is 1 bar (that is, $a(O_2) = p(O_2)/p^\ominus = 1$), equation 92 becomes

$$E = E^\ominus - \frac{RT}{2F}\ln\{a(H^+)\}^{-2} = E^\ominus + \frac{RT}{2F}\ln\{a(H^+)\}^2 = E^\ominus + \frac{RT}{F}\ln a(H^+)$$

Converting to logarithms to the base ten, (remember, $\ln x = 2.303 \log x$), substituting constants (at 298.15 K), and using the following more precise **definition of pH**,

$$pH = -\log a(H^+)\quad \text{(by definition)} \tag{93}$$

the variation of the potential with the hydrogen ion activity then reduces to:

$$E/V = 1.23 - 0.059\,2\ pH \tag{94}$$

- At what pH does the electrode potential of this couple have its standard value, $E^\ominus = 1.23$ V?

For the couple in equation 91, standard conditions imply $a(H^+) = 1$; that is, pH = 0 according to the definition in equation 93 – a conclusion that agrees with the expression in equation 94.

Now, equation 94 can be interpreted as a measure of the 'oxidizing power' of oxygen in a moist environment of varying pH. Its immense practical importance stems from the fact that our polluted atmosphere provides just such an environment! As you will see in the Topic Study associated with this part of the Course, metals **corrode** by being oxidized. This is a complex process, but one of the 'driving reactions' that is certainly important (in the initial stages, at least) is the couple in equation 91. The following SAQ provides an opportunity for you to assess the 'stability' of one or two metals with respect to the process of corrosion.

SAQ 19 Use information from Table 5 to decide which of the following metals would be expected to corrode in contact with water made slightly acid (pH = 5 say): (a) iron; (b) aluminium; (c) tin. Take the criterion of corrosion to be the thermodynamic tendency to form a solution of ions (Fe^{2+}, Al^{3+} and Sn^{2+}, respectively) close to the metal surface, with a concentration of at least 10^{-6} mol dm^{-3}. Are your conclusions consistent with your everyday experience of the properties of these metals?

Couples that involve a sparingly soluble salt

In this context, the familiar ($Ag^+|Ag$) couple again provides a convenient example:

$$Ag^+(aq) + e = Ag(s);\ E^\ominus = +0.80\ \text{V} \qquad (10)$$

■ Use the Nernst equation to write an expression for the electrode potential of this couple.

■ $E = E^\ominus - (RT/F)\ln\{a(Ag)/a(Ag^+)\}$

$= E^\ominus - (RT/F)\ln\{a(Ag^+)\}^{-1}$, since $a(Ag) = 1$

$$E = E^\ominus + (RT/F)\ln a(Ag^+) \qquad (95)$$

Now concentrate on the 'thought experiment' depicted in Figure 28. Here, the starting point (Figure 28a) comprises silver metal immersed in an aqueous solution containing unit activity of $Ag^+(aq)$ – a situation that represents the ($Ag^+|Ag$) couple set up under standard conditions. Measured relative to the S.H.E., the electrode potential of this system is just $E^\ominus(Ag^+|Ag)$.

Next, imagine that sulfide ions, $S^{2-}(aq)$, are added to the solution (Figure 28b). The immediate effect will be to precipitate out a black solid: this is the sparingly soluble salt, silver(I) sulfide (Ag_2S). Because the formation of Ag_2S *reduces* the activity of silver ions in the solution, this changes the electrode potential of the ($Ag^+|Ag$) couple (equation 95). The effect can be quantified by recalling the discussion of standard solubility products in Section 4.7.1. Specifically, the standard solubility product of Ag_2S controls the activities of $Ag^+(aq)$ and $S^{2-}(aq)$ that can coexist in *any* aqueous solution, as:

$$Ag_2S(s) = 2Ag^+(aq) + S^{2-}(aq)$$

$$K^\ominus_{sp} = \frac{\{a(Ag^+)\}^2 a(S^{2-})}{a(Ag_2S)} \quad \text{and } a(Ag_2S) = 1$$

So $K^\ominus_{sp} = \{a(Ag^+)\}^2 a(S^{2-}) = 6.62 \times 10^{-50}$ (at 298.15 K)

Figure 28 A representation of the 'experiment' described in the text.

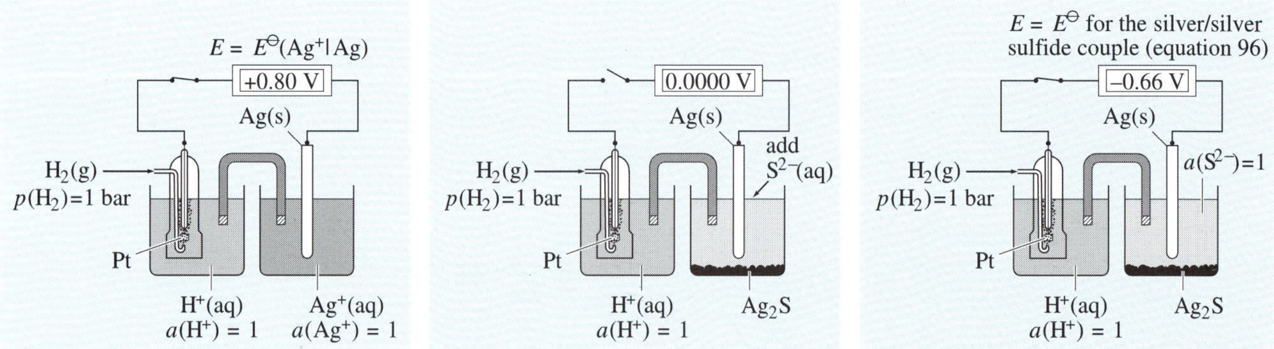

(a) The ($Ag^+|Ag$) couple set up under standard conditions; measured relative to the S.H.E. $E(Ag^+|Ag) = E^\ominus(Ag^+|Ag)$.

(b) $S^{2-}(aq)$ is added to the $Ag^+(aq)$ solution; a precipitate of solid Ag_2S is formed.

(c) In the final solution, $a(S^{2-}) = 1$, and $a(Ag^+)$ is determined by the standard solubility product of Ag_2S. Now $E(Ag^+|Ag)$ is the value of $E^\ominus$ for the couple in equation 96.

Suppose now that at the end of our 'experiment' (Figure 28c), the solution contains *unit activity* of sulfide ions.

■ What is the value of $E(Ag^+|Ag)$ under these circumstances?

■ $E(Ag^+|Ag) = -0.66$ V. If $a(S^{2-}) = 1$, then $a(Ag^+) = (K_{sp}^{\ominus})^{1/2} = 2.57 \times 10^{-25}$. With $E^{\ominus} = +0.80$ V, equation 95 becomes

$$E = \{+0.80\ \text{V} + 0.0257\ \text{V} \ln(2.57 \times 10^{-25})\}$$

$$= (+0.80 - 1.455)\ \text{V} = -0.66\ \text{V}$$

The analysis above is not just an exercise in the use of the Nernst equation and standard solubility products. Rather, it serves to illustrate one route to values of the *standard* electrode potential for couples that involve a sparingly soluble salt. In this particular case, the value of $E(Ag^+|Ag)$ calculated above is recorded in the literature as the standard electrode potential of the following couple (see Table 6):

$$Ag_2S(s) + 2e = 2Ag(s) + S^{2-}(aq);\ E^{\ominus} = -0.66\ \text{V} \qquad (96)$$

Notice that for this couple, the expression 'standard conditions' *implies unit activity of sulfide ions*. That is the crucial point to appreciate. The value of $E^{\ominus}$ is then taken to be the value of E for the $(Ag^+|Ag)$ couple when $a(S^{2-}) = 1$, such that the activity of silver ions is determined by the standard solubility product of Ag_2S.

The analysis outlined above is quite general: it can be applied equally well to other metal/insoluble salt couples. You may like to check that you have grasped the key point by working it through for the silver/silver chloride electrode – and there is another, slightly different, example in SAQ 24 (Section 6.3).

One last point. The example discussed here was chosen to bring the 'story' begun in Section 3 'full circle': the following SAQ invites you to pursue that idea. Taking a broader perspective, this sort of analysis has profound implications for the redox behaviour of metals and their aqueous ions in different environments. These implications are explored in the third-level course in inorganic chemistry (in the context of transitional metal chemistry) – and touched on in Topic Study 3 (in the context of metallic corrosion).

SAQ 20 Are the values of $E^{\ominus}$ for the couples in equations 10 and 96 consistent with the experimental behaviour of the cell discussed in Section 3 (and demonstrated in the associated video sequence)? Include in your answer any reservations you would have about drawing firm conclusions on the basis of the values of $E^{\ominus}$.

6.3 Summary of Sections 5 and 6

In the last two Sections we have succeeded in placing our treatment of electrochemical cells on a quantitative footing. This has enabled us both to rationalize the qualitative insights developed in Sections 2 and 3, and to establish the link between the cell emf E and a more general measure of 'reaction tendency'; that is, $(dG/d\xi) = -nFE$. This, in turn, has given you access to an alternative, and rather convenient, source of such data in the chemical literature – standard electrode potentials.

Provided only that the requisite data are available, you are now in a position to calculate $E^{\ominus}$ or, with the help of the Nernst equation, E for any cell or reaction you care to write down. The more important points are collected in Box 6, and tested in SAQs 15–20 above, and 21–25 below: make sure you try them at some stage.

Box 6 Using standard electrode potential data

1 For an electrochemical cell represented by a particular cell diagram, write the implied cell reaction as the *difference* between two reduction processes, RHE – LHE, each involving the gain of n electrons. Handle a 'normal' redox reaction by 'decomposing' it in the same way.

2 *Under standard conditions*:

$$E^{\ominus}\text{(overall)} = E^{\ominus}_{\text{RHE}} - E^{\ominus}_{\text{LHE}}$$
$$= (RT/nF)\ln K^{\ominus} \{= -\Delta G^{\ominus}_{\text{m}}/nF\}$$

If $E^{\ominus} > 0$ (or $\Delta G^{\ominus}_{\text{m}} < 0$), the reaction has a tendency to go left → right, as written.

If $E^{\ominus} < 0$ (or $\Delta G^{\ominus}_{\text{m}} > 0$), the reaction has a tendency to go left ← right, as written.

3 *Under non-standard conditions*:

$$E\text{(overall)} = E^{\ominus} - (RT/nF)\ln Q \quad \text{Nernst equation}$$

$$\text{cf. } (\text{d}G/\text{d}\xi) = \Delta G^{\ominus}_{\text{m}} + RT\ln Q$$

4 If $|E^{\ominus}| \geq 0.5$ V, then the sign of $E^{\ominus}$ is usually – *but not always* (see e.g. the example in SAQ 20) – a reliable indicator of the direction of spontaneous change.

SAQ 21 According to thermodynamic considerations alone, decide whether each of the following statements is true or false. Any information you require should be taken from Tables 5 and 6.

(a) If the following cell is set up under standard conditions at 298.15 K, then electrons should flow spontaneously from left to right through the external circuit:

$$\text{Ni(s)}|\text{Ni}^{2+}\text{(aq)}|\text{Al}^{3+}\text{(aq)}|\text{Al(s)}$$

(b) At 298.15 K, an acidified permanganate solution, MnO_4^-(aq), can be used to oxidize Sn^{2+}(aq) to Sn^{4+}(aq).

(c) In an aqueous solution of pH = 1, ferrous ions, Fe^{2+}(aq), are unstable to oxidation to ferric ions, Fe^{3+}(aq), by atmospheric oxygen.

SAQ 22 Consider the electrochemical cell represented by the following cell diagram:

$$\text{Ag(s)}|\text{Ag}^{+}\text{(aq)}|\text{Ni}^{2+}\text{(aq)}|\text{Ni(s)}$$

in which the concentrations of Ag^+(aq) and Ni^{2+}(aq) are 0.1 mol dm^{-3} and 1.0 mol dm^{-3}, respectively.

(a) What is the implied cell reaction?

(b) Use information from Table 5 to calculate the emf of the cell at 298.15 K, under the conditions specified above. State any assumptions or approximations involved in your calculation. What is the spontaneous cell reaction under these circumstances?

SAQ 23 Show that the pH-dependence of the electrode potential of the ($H^+|H_2$) couple is given by the following expression at 298.15 K:

$$E/\text{V} = -0.059\,2\,\text{pH}$$

(Assume that $p(H_2) = p^{\ominus} = 1$ bar.)

SAQ 24 Given that the standard solubility product of zinc hydroxide, $Zn(OH)_2$, is 7.61×10^{-17} (at 298.15 K), determine the standard electrode potential of the following couple:

$$Zn(OH)_2(s) + 2e = Zn(s) + 2OH^-(aq)$$

SAQ 25 One of the newer types of battery under investigation is called a 'mercury dry cell'. It involves combining the couple in SAQ 24 with the following half-reaction (at 298.15 K):

$$HgO(s) + H_2O(l) + 2e = Hg(l) + 2OH^-(aq); \quad E^{\ominus} = +0.098\ V$$

Determine the standard emf of the cell at 298.15 K and the maximum amount of electrical work obtainable under these conditions. Which electrode would be the anode and which the cathode? Would you expect the emf of the cell to depend on the pH of the electrolyte?

7 THERMODYNAMIC DATA FROM ELECTROCHEMICAL CELLS

Having discussed the tabulation and use of standard electrode potentials, we turn now to the question of how such data can be obtained by direct electrochemical methods. As we hinted earlier, it is remarkable how few of the values of $E^{\ominus}$ listed in the chemical literature were, in fact, obtained in this way. Nevertheless, where such measurements are possible, we shall see that they, in turn, provide a valuable source of other types of thermodynamic data.

7.1 Problems with the measurement of cell emfs

There are two basic problems with the direct determination of an emf. The first, outlined in Section 2.2.1, lies with the electron-transfer reactions at the individual electrodes. Here, as in any chemical reaction, *kinetic factors* can hinder the rapid establishment of equilibrium at an electrode. It is then difficult to determine accurately the zero-current condition necessary for the measured potential to be equated with the cell emf. The role of kinetic factors in these, and other, electrochemical systems is spelt out more clearly in Section 8 and taken up in Block 8. For the moment, we simply note the existence of the problem and the fact that it renders many cells effectively useless for the accurate determination of emfs.

The second problem lies with the construction of the cell, and in particular with whether or not the two half-cells involved contain electrolytes of different composition. More often than not this is unavoidable: the trouble is that a small, but finite, potential difference – called the **liquid junction potential** – will then be established between the two solutions. For reasons that are not properly understood, the use of a salt bridge effectively ameliorates this problem. A partial explanation may be that the bridge introduces two junction potentials, and it is thought that these tend to cancel.

Under the most favourable circumstances, this second problem no longer arises, because there is no liquid junction. A simple, but very important, example is shown in Figure 29: it comprises a hydrogen electrode and a silver/silver chloride electrode, both in the *same* electrolyte, dilute hydrochloric acid. The importance of a set-up like this is that it provides a route to the determination of *standard* electrode potentials.

7.2 The determination of standard electrode potentials

Here, we use the cell in Figure 29 to illustrate the procedure for determining a standard electrode potential. First, a word about the cell diagram in the caption to Figure 29 is in order. To reinforce the point that the cell *is* made up of two half-cells, the following 'expanded' version is rather more illuminating:

Pt, $H_2(g)|H^+(aq)....Cl^-(aq)|AgCl(s)|Ag(s)$

The combination on the left is just a representation of the hydrogen electrode – familiar from earlier Sections. The combination on the right is the conventional way of representing the silver/silver chloride electrode – *when it forms the RHE*, that is. If it forms the LHE, the *order* is reversed: it then reads $Ag(s)|AgCl(s)|Cl^-(aq)$. This notation is used for *all* metal/insoluble salt electrodes. Although it may seem a little curious, notice that for the metal it does follow the rule (reduced state | oxidized state) for the LHE, and (oxidized state | reduced state) for the RHE: the novel feature is the inclusion of the anion involved.

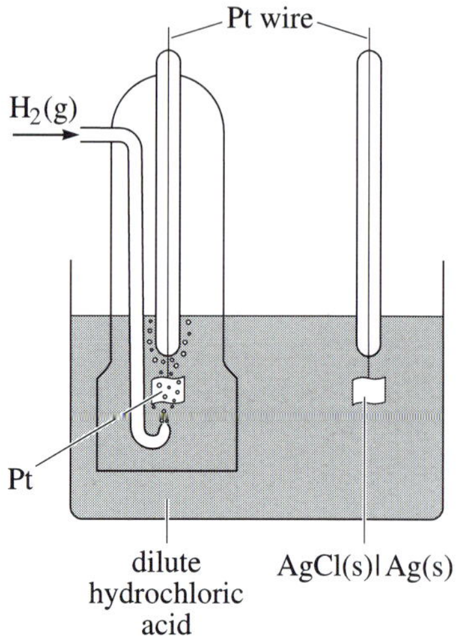

Figure 29 Experimental arrangement, without a liquid junction, for the cell: Pt, $H_2(g)|HCl(aq)|AgCl(s)|Ag(s)$.

- Write an equation for the implied cell reaction.

- This can be thought of as the *difference* (RHE – LHE) between the following reduction processes:

RHE:	$AgCl(s) + e = Ag(s) + Cl^-(aq)$	(86)
LHE:	$H^+(aq) + e = \frac{1}{2}H_2(g)$	(7)
RHE – LHE:	$\frac{1}{2}H_2(g) + AgCl(s) = H^+(aq) + Ag(s) + Cl^-(aq)$	(97)

- Now use the Nernst equation to write an expression for the emf of this cell. Remember that the activities of solids may be taken as unity, and assume that the hydrogen gas is at the standard pressure of 1 bar (so that $a(H_2) = p(H_2)/p^\ominus = 1$).

- With the provisos above, and noting that $n = 1$ in this case, the cell emf is given by:

$$E = E^\ominus - \frac{RT}{F}\ln\left(\frac{a(H^+)a(Ag)a(Cl^-)}{\{a(H_2)\}^{1/2}a(AgCl)}\right)$$

$$E = E^\ominus - (RT/F)\ln\{a(H^+)a(Cl^-)\} \quad (98)$$

The crucial point is that the standard emf of this *cell*, $E^\ominus$ in equation 98, is *defined* to be the standard electrode potential of the silver/silver chloride electrode (Section 6.1.1). How, then, does the expression in equation 98 help to determine this quantity?

You encountered a similar problem, essentially the determination of standard equilibrium constants, in Section 4.7.4. Refer back to that discussion. Can you now suggest a way of tackling the question posed above?

The secret again lies with an *extrapolation*, based on the Debye–Hückel limiting law. The first step is to rewrite equation 98 in terms of the concentration c of the acid (where $c = c(H^+) = c(Cl^-)$) and its mean ionic activity coefficient $\gamma_\pm = (\gamma_+\gamma_-)^{1/2}$:

$$E = E^\ominus - (RT/F)\ln\{(\gamma_+c/c^\ominus)(\gamma_-c/c^\ominus)\}$$

$$= E^\ominus - (RT/F)\ln(c/c^\ominus)^2 - (RT/F)\ln\gamma_\pm^2$$

or

$$E + 2(RT/F)\ln(c/c^\ominus) = E^\ominus - 2(RT/F)\ln\gamma_\pm \quad (99)$$

As you saw in Section 4.7.3, for a dilute 1 : 1 electrolyte with univalent ions, the ionic strength is equal to $(c/c^{\ominus})$ and the Debye–Hückel limiting law takes the form:

$$\log \gamma_{\pm} = -0.51(c/c^{\ominus})^{1/2}$$

or

$$\ln \gamma_{\pm} = -1.17(c/c^{\ominus})^{1/2}$$

so equation 99 becomes:

$$\{E + 2(RT/F)\ln(c/c^{\ominus})\} = E^{\ominus} + 2.34(RT/F)(c/c^{\ominus})^{1/2} \quad (100)$$

Notice how equation 100 is arranged: in particular, the expression on the left may be readily determined by measuring the emf E of the cell for a range of concentrations c of the acid it contains. As Figure 30 shows, the desired value of $E^{\ominus}$ is just the intercept (at $c = 0$) on a plot of this quantity against the square root of the (dimensionless) concentration. Once again, the limiting law neatly circumvents the *hypothetical* nature of the standard state, and thereby allows determination of a standard thermodynamic quantity.

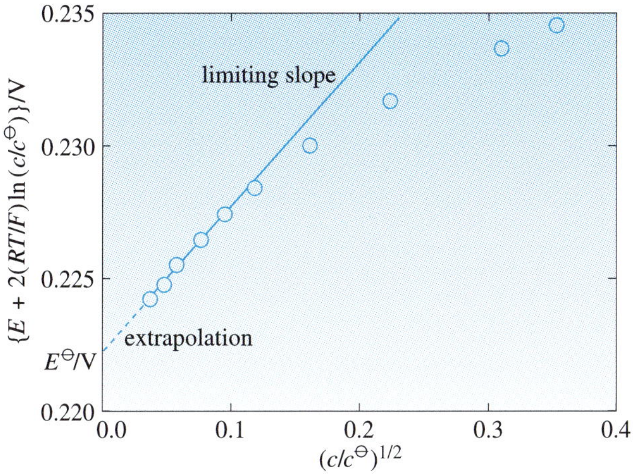

Figure 30 Experimental determination of the standard emf of the cell in Figure 29.

On a more pragmatic note, the example considered above derives additional importance from the widespread use of the silver/silver chloride electrode as a convenient **secondary** or **reference electrode** in the determination of other electrode potentials. The reasons are two-fold. First, the primary standard (the hydrogen electrode) is tricky to set up: unless handled with care, it often yields results that are difficult to reproduce. Second, the silver/silver chloride electrode is 'sensitive' to the activity of an anion. It can therefore share an electrolyte with a 'cation-sensitive' electrode (a metal, say), and the system is then amenable to the rather simple analysis outlined above.

Once the value of $E^{\ominus}$ (for a cell or an electrode) is known, it can of course be used to determine other thermodynamic quantities for the reaction in question. One example, the determination of equilibrium constants, was discussed in Section 6.2.1. Two further examples that you should be able to work through for yourself are contained in the following SAQs. (Notice the accuracy required if cells are to be used as a source of thermodynamic data.)

SAQ 26 From Figure 30, $E^{\ominus}(Cl^-|AgCl|Ag) = 0.222\,3$ V at 298.15 K. Given that $\Delta G_f^{\ominus}$ (AgCl, s) = −109.8 kJ mol^{-1}, determine the standard Gibbs function of formation of Cl^-(aq); that is, the value of $\Delta G_f^{\ominus}$ (Cl^-, aq) at 298.15 K.

SAQ 27 The emf of the cell discussed in this Section is 468.6 mV when the HCl is at a concentration of 9.14×10^{-3} mol dm^{-3}. Using the value of $E^{\ominus}(Cl^-|AgCl|Ag)$ given in SAQ 26, calculate the mean ionic activity coefficient of hydrochloric acid of this concentration. [*Hint* Use equation 99.]

7.3 The temperature-dependence of the emf

So far, we have concentrated exclusively on the determination, and use, of electrochemical data at *one* temperature, 298.15 K. We turn now to the temperature-dependence of such data. It is here that electrochemical cells can provide a further valuable contribution to the stores of thermodynamic data in the literature. How, then, does $E^{\ominus}$ change with temperature?

Well, for any reaction at *any* (constant) temperature T,

$$\Delta G_{\mathrm{m}}^{\ominus} = \Delta H_{\mathrm{m}}^{\ominus} - T\Delta S_{\mathrm{m}}^{\ominus}$$

which, in view of the relation between $\Delta G_{\mathrm{m}}^{\ominus}$ and $E^{\ominus}$ ($\Delta G_{\mathrm{m}}^{\ominus} = -nFE^{\ominus}$, equation 80) can be written as follows:

$$-nFE^{\ominus} = \Delta H_{\mathrm{m}}^{\ominus} - T\Delta S_{\mathrm{m}}^{\ominus} \qquad (101)$$

While equation 101 effectively answers the question posed above, its importance is revealed most clearly by recalling an *approximation* that was introduced in the Second Level Inorganic Course, and discussed further in Block 1 of this Course:

> Provided the physical states of the substances do not change, then it is usually a good approximation to assume that the values of $\Delta H_{\mathrm{m}}^{\ominus}$ and $\Delta S_{\mathrm{m}}^{\ominus}$ for a reaction do not change with temperature.*

Suppose now that the standard emf of a cell is determined at two different temperatures, T_1 and T_2, say. From equation 101, and with the assumption above,

$$-nFE_1^{\ominus} = \Delta H_{\mathrm{m}}^{\ominus} - T_1\Delta S_{\mathrm{m}}^{\ominus}$$

and

$$-nFE_2^{\ominus} = \Delta H_{\mathrm{m}}^{\ominus} - T_2\Delta S_{\mathrm{m}}^{\ominus}$$

It then follows that the change in $E^{\ominus}$ is given by

$$E_2^{\ominus} - E_1^{\ominus} = \Delta S_{\mathrm{m}}^{\ominus}(T_2 - T_1)/nF \qquad (102)$$

In other words, the temperature-dependence of $E^{\ominus}$ gives the standard molar *entropy* change, $\Delta S_{\mathrm{m}}^{\ominus}$, for the underlying cell reaction. Because the majority of cell reactions involve aqueous ions, this in turn can provide a valuable, non-calorimetric method for determining the *absolute entropies* $S^{\ominus}$ of such species.

- Do you recall the *arbitrary* convention implicit in the assignment of $S^{\ominus}$ values to individual aqueous ions?

- As with the assignment of $\Delta H_{\mathrm{f}}^{\ominus}$, $\Delta G_{\mathrm{f}}^{\ominus}$ and $E^{\ominus}$ values, the value of $S^{\ominus}$ for the hydrogen ion is chosen as an arbitrary zero:

$$S^{\ominus}(\mathrm{H}^+, \mathrm{aq}) = 0 \qquad \text{(by definition)}$$

SAQ 28 At 20 °C, $E^{\ominus}(\mathrm{Cl}^-|\mathrm{AgCl}|\mathrm{Ag})$ is 0.225 6 V, and at 30 °C it is 0.219 1 V. (Again, notice the accuracy required for these emf measurements.) Find the values of $\Delta S_{\mathrm{m}}^{\ominus}$ and $\Delta H_{\mathrm{m}}^{\ominus}$ for the electrode reaction, and of $S^{\ominus}$ for the $\mathrm{Cl}^-(\mathrm{aq})$ ion, all at 298.15 K. Any further information you require should be taken from your S342 *Data Book*. (Warning! There is a trap for the unwary in this question. Make sure you write out the *full* reaction to which the value of $E^{\ominus}(\mathrm{Cl}^-|\mathrm{AgCl}|\mathrm{Ag})$ refers: see Section 7.2).

* In fact, $\Delta H_{\mathrm{m}}^{\ominus}$ and $\Delta S_{\mathrm{m}}^{\ominus}$ sometimes change surprisingly much with temperature for 'ionic' reactions. Nevertheless, this is still a reasonable approximation *provided* the temperature range is restricted to a *few* degrees.

7.4 Summary of Section 7

By this stage, it should be clear that the various types of thermodynamic data recorded in the literature are intimately interrelated. As far as electrode potentials are concerned, relatively few are ultimately determined by the direct electrical measurements described here: few electrode systems are free from the kinetic problems mentioned in Section 7.1. The most important alternative source of such data is calorimetry, experiments that yield enthalpy and entropy data. Conversely, these data can, in turn, be derived from experimental emfs where such measurements are possible.

SAQ 29 Use information from the S342 *Data Book* to calculate $\Delta S_{\mathrm{m}}^{\ominus}$ at 298.15 K for the reaction in a car battery:

$$Pb(s) + PbO_2(s) + 4H^+(aq) + 2SO_4^{2-}(aq) = 2PbSO_4(s) + 2H_2O(l) \tag{89}$$

Given that $E^{\ominus} = 2.05$ V at 25 °C, calculate the corresponding value at 0 °C. To what extent does your answer explain the difficulties that may be experienced in starting a car on a winter morning?

8 ELECTROCHEMISTRY IN ACTION:

LIMITATIONS OF THE THERMODYNAMIC APPROACH

As you saw in Section 6, electrode reactions can be arranged in order of their $E^{\ominus}$ values such that, under standard conditions, the oxidized state of a particular couple is thermodynamically capable of oxidizing the reduced state of any couple below it. This statement includes two very important qualifications. The first of these, implicit in the phrase 'under standard conditions', was discussed in Section 6.2. You saw there that the Nernst equation effectively lifts this restriction, and hence allows similar predictions to be made under non-standard conditions. Again, the discussion in Section 7.3 suggests that this procedure could, if the need arose, be extended to temperatures other than 298.15 K.

We come now to the second qualification: like all thermodynamic data, electrode potentials are subject to the limitations imposed by kinetics. You have already met one or two examples. Thus, despite the prediction you made in answering SAQ 16, no noticeable reaction takes place when a piece of aluminium metal is dropped into copper sulfate solution. In this case, the '*kinetic stability*' of the system can be attributed to a thin, but adherent, film of oxide (Al_2O_3) on the surface of the aluminium. Indeed, it is an oxide layer like this that protects many of the metals in everyday use from the immensely destructive corrosion processes that should, according to thermodynamics alone, return them to the 'oxidized' state. (See the answer to SAQ 19.)

Unhappily, one of the most important structural metals, iron, appears to be largely devoid of such 'intrinsic' protection. For a particular structure made of iron, interest then centres not only around the 'tendency for corrosion' in a specified environment (the realm of thermodynamics), but also, and more importantly, around the *rate* of that process. It is this second factor that determines the useful lifetime of the structure in question, be it a car, a chemical plant or a drilling platform in the North Sea.

The examples cited above simply serve to underline the importance of kinetic factors in determining the progress of *any* chemical reaction. But how do these factors manifest themselves when the reaction in question is an *electrochemical* process – that is, when it actually takes place via complementary electron-transfer reactions at *separated* sites? As you will see in Topic Study 3, in many cases corrosion is, in fact, an electrochemical process in this sense.

To examine this question, we return now to the subject with which we began our study of electrochemistry: electrolysis.

8.1 The electrochemical 'substance producer': electrolysis

So far, we have concentrated on electrochemical cells in which the underlying cell reaction is a spontaneous process. As an example, consider again the lead accumulator and its cell reaction:

$$Pb(s) + PbO_2(s) + 4H^+(aq) + 2SO_4^{2-}(aq) = 2PbSO_4(s) + 2H_2O(l) \quad (89)$$

Suppose that the condition of the electrolyte is such that the cell emf is 1.89 V (as calculated in Section 6.2.2): under these conditions, the emf of the *reaction* in equation 89 is 1.89 V. Conversely, the emf of the reverse reaction is −1.89 V: it is a non-spontaneous process. Nevertheless, the reaction can be 'driven' in this direction by the simple expedient of applying an *opposing* potential greater than 1.89 V. This is how a car battery is 'recharged'.

Precisely the same analysis can be applied to any non-spontaneous process: the numerical value of its (by definition negative) emf then represents the *minimum* potential necessary to drive the reaction. This is the essence of electrolysis.

Take the electrolysis of water as a familiar example:

$$H_2O(l) = H_2(g) + \tfrac{1}{2}O_2(g) \quad (6)$$

To summarize the results of SAQ 1 (Section 1.1): with the set-up shown schematically in Figure 3 (repeated here as Figure 31), the half-reactions at the two electrodes can be represented as follows:

anode (oxidation): $H_2O(l) = \tfrac{1}{2}O_2(g) + 2H^+(aq) + 2e$

cathode (reduction): $2H^+(aq) + 2e = H_2(g)$

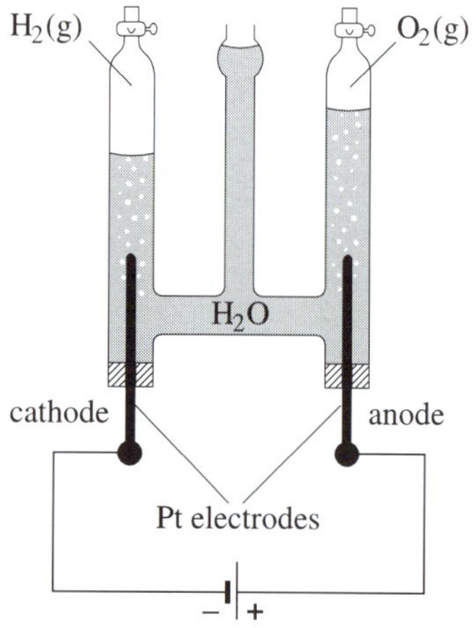

Figure 31 The electrolysis of water (small amount of acid added).

Alternatively – and to be consistent with the discussion in Section 6 – the overall reaction in equation 6 can be written as the *difference* (RHE – LHE) between the following couples, *both written as reductions*:

RHE: $2H^+(aq) + 2e = H_2(g)$; $E/V = -0.0592\,pH$

LHE: $\tfrac{1}{2}O_2(g) + 2H^+(aq) + 2e = H_2O(l)$; $E/V = 1.23 - 0.0592\,pH$

where the pH-dependence of the electrode potentials comes from the analysis (using the Nernst equation) in Section 6.2.3 (see equation 94 and SAQ 23).

- On this basis, what is the minimum potential necessary to achieve the electrolysis in equation 6?

- For equation 6,

$$E = E_{RHE} - E_{LHE} = (-0.0592\,pH)\,V - (1.23 - 0.0592\,pH)\,V = -1.23\,V.$$

Thus, to produce hydrogen and oxygen from water, a potential of at least 1.23 V must be applied between the electrodes of the cell – a value that is *independent* of the pH of the solution.

Look back at the discussion in Section 5.2. What is the implication of the above analysis from a *kinetic* point of view?

The minimum potential, as calculated from emf values, is a thermodynamic quantity: as such, it corresponds to a situation in which an infinitesimal current passes through the cell. Now, current is just a measure of the *rate* at which electrons flow, both through the external circuit and, thence, across the electrode/electrolyte interfaces at which the half-reactions occur. (The SI unit of current – the amp, A – reflects this connection: $1\,A = 1\,C\,s^{-1}$, that is, the rate of charge flow.) Thus, with an infinitesimal current, the overall reaction (equation 6, say) is brought about extremely slowly!

It follows that to produce a substance (hydrogen, say, in the example above) at an acceptable rate, the current passing through the cell must be increased. In practice, this is achieved by raising the potential *above* the minimum value, as we hinted above. The crucial point is that the effect of this increase on the rate (or even the outcome) of a particular electrolysis *cannot* be predicted on the basis of thermodynamic arguments alone. This point is made quite forcibly by the following example.

Suppose that the water in Figure 31 is replaced by an acidified solution of zinc sulfate: roughly 50% of zinc metal is, in fact, produced by the electrolysis of a solution like this.

SAQ 30 Under industrial conditions, the electrolyte contains $Zn^{2+}(aq)$ at a concentration of roughly 2.0 mol dm^{-3} and its pH is about 3. Assuming that O_2 is liberated at the anode, what is the minimum potential necessary: (a) to deposit zinc metal at the cathode, and (b) to liberate H_2 at the cathode, under these conditions? State any assumptions involved in your calculations. Do you foresee a problem?

To achieve a reasonable throughput, a potential of around 3.5 V is used in practice. Under these circumstances, the thermodynamic analysis in SAQ 30 suggests that *both* hydrogen evolution *and* zinc deposition will occur at the cathode. With platinum electrodes, this is indeed the case, with hydrogen evolution predominating. But *if the cathode material is changed* (to zinc or aluminium, for example), the situation is reversed: *zinc is deposited in preference to hydrogen*. The analysis in SAQ 30 contains no hint of this possibility, yet it is clearly of vital importance to the zinc production industry.

In view of our earlier comments, it seems likely that the explanation for this behaviour lies with the *rates* of the two electron-transfer processes that could, in principle at least, take place at the cathode. Further, the example above suggests strongly that the rate of the hydrogen-evolution reaction depends not only on the applied potential, but also on the *nature* of the surface (electrode) at which it takes place. In what ways must the kinetic treatment developed in previous Blocks be modified in order to take account of factors like this? Again, we have implied a link between the current flow and the rate of an electrode reaction, but what is the precise nature of this link?

It is these and other similar questions that we shall examine in the next Block. We shall then be in a position to return to practical electrochemical systems, and examine the interplay of thermodynamic *and* kinetic factors in their design and efficient operation.

OBJECTIVES FOR BLOCK 7

Now that you have completed Block 7, you should be able to do the following things:

1 Recognize valid definitions of, and use in a correct context, the terms, concepts and principles printed in bold type in the text and collected in the following Table.

List of scientific terms, concepts and principles used in Block 7

Term	Page No.	Term	Page No.
activity, a	25	maximum electrical work, $w_{el,max}$	41
activity coefficient, γ	32	mean ionic activity coefficient, $\gamma_{\pm}$	32
anode	7, 10	metal/insoluble salt electrode	49, 54
calculating 'reaction tendency', $(dG/d\xi)$	27	Nernst equation	44
cathode	7, 10	pH	53
cell diagram	13	reaction quotient, Q	29
chemical potential, μ	23	redox couple (or half-reaction)	8, 45
corrosion	53	reference electrode	59
Debye–Hückel limiting law	35	relation between E and $(dG/d\xi)$	43
'deviations' from ideality	32	reversible change	41
electrochemical cell	9	salt bridge	9
electrochemical reaction	8	saturated solution	30
electrode	6	self-driving cell	8
electrolysis	6, 62	sign convention for emf	15
electrolytic cell	6	solubility product, K_{sp}	31
electromotive force (emf), E	11	standard chemical potential, $\mu^{\ominus}$	25
electron-transfer reaction	8	standard conditions	25
extent of reaction, ξ	21	standard electrode potential	46
half-cell	9	standard emf, $E^{\ominus}$	44
ideal behaviour	26	standard equilibrium constant, $K^{\ominus}$	29
implied cell reaction	13	standard hydrogen electrode (S.H.E.)	46
infinite dilution	36	standard molar Gibbs function change, $\Delta G_m^{\ominus}$	26
ionic strength, I	35	standard solubility product, $K_{sp}^{\ominus}$	31
liquid junction potential	57	standard state	25
		temperature-dependence of emf	60

2 Given the cell diagram for an electrochemical cell, write down the implied cell reaction. (SAQs 3, 4, 21, 22 and 25)

3 From the cell diagram and a knowledge of the cell polarity, attach a sign to the cell emf E, and predict the spontaneous cell reaction. (SAQs 3, 4 and 14)

4 Define the activity of: (a) a solid or a liquid; (b) an ideal gas; (c) a solute in an 'ideal' solution; (d) a solute in a 'non-ideal' solution; and (e) an aqueous electrolyte. (SAQs 7–9, 12, 19, 20, 22–24, 30; Exercise 2)

5 Given a balanced reaction equation, use the definitions in Objective 4 to write down an expression for the 'reaction tendency' $(dG/d\xi)$ in terms of $\Delta G_m^{\ominus}$ and the composition of the system, and:

(a) calculate the value of $(dG/d\xi)$ under stated conditions, and hence predict the direction of spontaneous change;

(b) state any assumptions or approximations involved in this calculation. (SAQs 6–8, and 14; Exercise 2)

6 Write down the standard equilbrium constant $K^{\ominus}$ for a given reaction, and use the definitions in Objective 4 to relate this to the corresponding 'experimental' equilibrium constant, expressed in terms of: (a) partial pressures, K_p; (b) concentrations, K_c and K_{sp}; or (c) both. (SAQs 5, 9, 12 and 24; Exercise 2)

7 Given appropriate information, use the relation in Objective 6b to determine the mean ionic activity coefficient $\gamma_{\pm}$ of a sparingly soluble electrolyte. (SAQs 9 and 12)

8 Given the composition of an electrolyte solution, calculate its ionic strength. (SAQs 10 and 12)

9 Describe briefly the physical basis of the Debye–Hückel theory of ionic solutions, and use the Debye–Hückel limiting law:

(a) to explain observed deviations from ideal behaviour in electrolyte solutions; (SAQs 10, 11 and 12)

(b) to determine the values of standard thermodynamic quantities. (SAQ 11)

10 Determine the maximum electrical work obtainable from a reaction, under stated conditions. (SAQs 13 and 25)

11 Use appropriate values of $\Delta G_{\mathrm{m}}^{\ominus}$ to calculate:

(a) the standard emf $E^{\ominus}$ of a given reaction or cell reaction (or vice versa);

(b) the standard electrode potential of a given redox couple (or vice versa). (SAQs 15 and 26)

12 Use standard electrode potential data to:

(a) calculate the standard emf of a given electrochemical cell or reaction; (SAQs 17, 18, 21, 22 and 25)

(b) predict the likely course of a given reaction, under standard conditions; (SAQs 16, 17, 21 and 22)

(c) determine the value of $K^{\ominus}$ for a reaction (or vice versa). (SAQs 18 and 24)

13 Use the Nernst equation to determine the 'concentration'-dependence of individual electrode potentials or cell emfs, and hence:

(a) calculate the emf of a given electrochemical cell, under non-standard conditions;

(b) predict the likely course of a given reaction, under non-standard conditions;

(c) predict the likely products of electrolysis reactions in aqueous solution. (SAQs 19–24 and 30)

14 State the assumptions involved in the predictions outlined in Objectives 12 and 13, and their limitations. (SAQs 16, 19–22, 29 and 30)

15 Outline, without giving experimental detail, the use of electrochemical cells to determine a variety of thermodynamic data. (SAQs 26, 27 and 28)

16 Relate the temperature-dependence of $E^{\ominus}$ to the standard entropy change for the corresponding cell reaction. (SAQs 28 and 29)

SAQ ANSWERS AND COMMENTS

SAQ 1 (Objective 1)

(a) The addition of acid effectively raises the concentration of $H^+(aq)$ ions in the solution, and hence increases its conductivity. (In practice, this could be achieved equally well by adding hydroxide ions, but then the half-reactions at the two electrodes would be different from those given here.)

(b) From the hint in the question, the half-reaction at the positive electrode can be obtained by writing the overall reaction (equation 6), and then *subtracting* from this the half-reaction at the negative electrode, as

overall: $H_2O(l) = H_2(g) + \frac{1}{2}O_2(g)$

negative electrode: $2H^+(aq) + 2e = H_2(g)$

subtract: $H_2O(l) - 2H^+(aq) - 2e = \frac{1}{2}O_2(g)$

which can be reorganized to read:

positive electrode: $H_2O(l) = \frac{1}{2}O_2(g) + 2H^+(aq) + 2e$

As expected, this represents an oxidation process: water molecules are 'de-electronated' at the positive electrode, and electrons are thereby fed back into the external circuit.

(c) The connections to the external source, *and the processes involved*, identify the right- and left-hand electrodes as the anode (*oxidation*) and cathode (*reduction*), respectively.

SAQ 2 (Objective 1)

If copper and silver rods were simply dipped into a *common* electrolyte, containing $Ag^+(aq)$ ions, then the *chemical* reaction in equation 8 could take place directly. The copper could then reduce the silver ions without any transfer of electrons through the external circuit. The combination of two half-cells effectively prevents this 'short-circuit', and ensures that the reaction follows the *electrochemical* path.

SAQ 3 (Objectives 2 and 3)

(a) The arrangement in Figure 9, and the nature of the cell components, suggests the following cell diagram:

$$Pb(s)|Pb^{2+}(aq)|Sn^{2+}(aq)|Sn(s)$$

with (according to Box 2) the following implied cell reaction:

LHE (oxidation): $Pb(s) = Pb^{2+}(aq) + 2e$

RHE (reduction): $Sn^{2+}(aq) + 2e = Sn(s)$

overall: $Pb(s) + Sn^{2+}(aq) = Pb^{2+}(aq) + Sn(s)$

If you wrote the cell diagram the other way round, with Sn as the LHE, then the implied cell reaction would also be reversed.

(b) In the cell diagram in part (a), Sn is the RHE. If it is found to be the anode, the emf will be reported as negative, $E = -0.010$ V, and the implied cell reaction will have a spontaneous tendency to go left ← right as written above (that is, Sn is oxidized to Sn^{2+}, and Pb^{2+} is reduced to Pb). Again, if you wrote the cell diagram the other way round, then $E = +0.010$ V, and *your* implied cell reaction would go left → right, as written.

SAQ 4 (Objectives 2 and 3)

(a) The implied cell reaction is:

LHE (oxidation): $Zn(s) = Zn^{2+}(aq) + 2e$

RHE (reduction): $Cu^{2+}(aq) + 2e = Cu(s)$

overall: $Zn(s) + Cu^{2+}(aq) = Zn^{2+}(aq) + Cu(s)$

(b) With Zn (LHE) as the anode, $E > 0$ and the implied cell reaction should go left → right as written.

(c) For the reaction above,

$$\Delta G_m^\ominus = \Delta G_f^\ominus(Zn^{2+}, aq) + \Delta G_f^\ominus(Cu, s) - \Delta G_f^\ominus(Zn, s) - \Delta G_f^\ominus(Cu^{2+}, aq)$$

$$= \{(-147.1) + (0) - (0) - (65.5)\}\ \text{kJ mol}^{-1} = -212.6\ \text{kJ mol}^{-1}$$

The negative value of $\Delta G_m^\ominus$ confirms that the cell reaction, as written, should be a spontaneous process – or at least, it should under standard conditions at 298.15 K. More on this in the next Section.

SAQ 5 (Objective 6)

From equation 17,

$$\ln K^\ominus = -\Delta G_m^\ominus / RT = \frac{-(-5 \times 10^3\ \text{J mol}^{-1})}{(8.314\ \text{J K}^{-1}\ \text{mol}^{-1}) \times (298.15\ \text{K})} = +2.017$$

So $K^\ominus = 7.52$ (i.e. $K^\ominus > 1$).

In this case, the equilibrium constant is simply the ratio $(c_B/c_A)_e$ of the concentrations of A and B at *equilibrium*. Now, if $c_B = 100c_A$, then the initial mixture has a concentration ratio $c_B/c_A = 100$. For this simple system, the only way of achieving the equilibrium ratio (7.52) is if c_A increases at the expense of c_B. This corresponds to a spontaneous change from B to A, that is, in the *reverse* direction to that predicted on the basis of $\Delta G_m^\ominus$ alone.

SAQ 6 (Objective 5)

In each case, the expression for $(dG/d\xi)$ depends on the values of v in the balanced reaction equation *as written*. From equation 35:

(a) $(dG/d\xi) = \mu(H^+) + \mu(Ag) - \frac{1}{2}\mu(H_2) - \mu(Ag^+)$

(b) $(dG/d\xi) = \mu(Zn^{2+}) + \mu(Cu) - \mu(Zn) - \mu(Cu^{2+})$

(c) $(dG/d\xi) = \mu(Ag^+) + \mu(Cl^-) - \mu(AgCl)$

Notice that an expression for $(dG/d\xi)$ can be written down *only* if the reaction equation is specified.

SAQ 7 (Objectives 4 and 5)

Following the steps in Box 5,

$$
\begin{aligned}
(\mathrm{d}G/\mathrm{d}\xi) &= \mu_\mathrm{B} - \mu_\mathrm{A} \\
&= (\mu_\mathrm{B}^\ominus + RT\ln a_\mathrm{B}) - (\mu_\mathrm{A}^\ominus + RT\ln a_\mathrm{A}) \\
&= (\mu_\mathrm{B}^\ominus - \mu_\mathrm{A}^\ominus) + RT(\ln a_\mathrm{B} - \ln a_\mathrm{A}) \\
&= (\mu_\mathrm{B}^\ominus - \mu_\mathrm{A}^\ominus) + RT\ln(a_\mathrm{B}/a_\mathrm{A}) \\
&= \Delta G_\mathrm{m}^\ominus + RT\ln(a_\mathrm{B}/a_\mathrm{A})
\end{aligned}
$$

If A and B are solutes in an 'ideal' solution, then $a_\mathrm{B} = c_\mathrm{B}/c^\ominus$ and $a_\mathrm{A} = c_\mathrm{A}/c^\ominus$, and the expression above becomes:

$$
\begin{aligned}
(\mathrm{d}G/\mathrm{d}\xi) &= \Delta G_\mathrm{m}^\ominus + RT\ln(c_\mathrm{B}/c_\mathrm{A}) \\
&= (-5\,\mathrm{kJ\,mol^{-1}}) + (8.314\,\mathrm{J\,K^{-1}\,mol^{-1}} \times 298.15\,\mathrm{K})\ln(100) \\
&= (-5.0 + 11.42)\,\mathrm{kJ\,mol^{-1}} = +6.42\,\mathrm{kJ\,mol^{-1}}
\end{aligned}
$$

Thus, for these concentrations, $(\mathrm{d}G/\mathrm{d}\xi) > 0$ for the reaction A = B. This implies that the *reverse reaction*, for which $(\mathrm{d}G/\mathrm{d}\xi) < 0$, will be the spontaneous process under these conditions, as you found in answering SAQ 5.

SAQ 8 (Objectives 4 and 5)

(a) From Box 5, for reaction 16,

$$
\begin{aligned}
(\mathrm{d}G/\mathrm{d}\xi) &= \mu(\mathrm{H^+}) + \mu(\mathrm{Ag}) - \tfrac{1}{2}\mu(\mathrm{H_2}) - \mu(\mathrm{Ag^+}) \\
&= \{\mu^\ominus(\mathrm{H^+}) + RT\ln a(\mathrm{H^+})\} + \{\mu^\ominus(\mathrm{Ag}) + RT\ln a(\mathrm{Ag})\} - \\
&\quad \tfrac{1}{2}\{\mu^\ominus(\mathrm{H_2}) + RT\ln a(\mathrm{H_2})\} - \{\mu^\ominus(\mathrm{Ag^+}) + RT\ln a(\mathrm{Ag^+})\} \\
&= \{\mu^\ominus(\mathrm{H^+}) + \mu^\ominus(\mathrm{Ag}) - \tfrac{1}{2}\mu^\ominus(\mathrm{H_2}) - \mu^\ominus(\mathrm{Ag^+})\} + \\
&\quad RT\{\ln a(\mathrm{H^+}) + \ln a(\mathrm{Ag}) - \tfrac{1}{2}\ln a(\mathrm{H_2}) - \ln a(\mathrm{Ag^+})\} \\
&= \Delta G_\mathrm{m}^\ominus + RT\ln\left(\frac{a(\mathrm{H^+})a(\mathrm{Ag})}{\{a(\mathrm{H_2})\}^{1/2}a(\mathrm{Ag^+})}\right)
\end{aligned}
$$

For reaction 16,

$$Q = \left(\frac{a(\mathrm{H^+})a(\mathrm{Ag})}{\{a(\mathrm{H_2})\}^{1/2}a(\mathrm{Ag^+})}\right)$$

Comparison of the expressions above confirms that the short cut implied by equation 51 would lead to the same expression for $(\mathrm{d}G/\mathrm{d}\xi)$.

(b) The definitions in Cases 1–3 (Section 4.4) allow the activities to be translated into measurable quantities as: Ag(s), $a(\mathrm{Ag}) = 1$; $\mathrm{H_2}$(g), $a(\mathrm{H_2}) = p(\mathrm{H_2})/p^\ominus$; $\mathrm{H^+}$(aq) and $\mathrm{Ag^+}$(aq), $a(\mathrm{H^+}) = c(\mathrm{H^+})/c^\ominus$ and $a(\mathrm{Ag^+}) = c(\mathrm{Ag^+})/c^\ominus$. Thus

$$
\begin{aligned}
(\mathrm{d}G/\mathrm{d}\xi) &= \Delta G_\mathrm{m}^\ominus + RT\ln\frac{\{c(\mathrm{H^+})/c^\ominus\}(1)}{\{p(\mathrm{H_2})/p^\ominus\}^{1/2}\{c(\mathrm{Ag^+})/c^\ominus\}} \\
&= \Delta G_\mathrm{m}^\ominus + RT\ln\{c(\mathrm{H^+})/c(\mathrm{Ag^+})\} \qquad (\text{with } p(\mathrm{H_2}) = p^\ominus = 1\,\text{bar})
\end{aligned}
$$

Assumptions: $\mathrm{H_2}$ behaves as an ideal gas (at 1 bar) and both $\mathrm{H^+}$(aq) and $\mathrm{Ag^+}$(aq) behave as solutes in an ideal solution (at the concentrations in (i) and (ii) below).

(i) $$
\begin{aligned}
(\mathrm{d}G/\mathrm{d}\xi) &= (-77.1\,\mathrm{kJ\,mol^{-1}}) + (8.314 \times 10^{-3}\,\mathrm{kJ\,K^{-1}\,mol^{-1}} \times 298.15\,\mathrm{K})\ln(10^{-5}/10^{-1}) \\
&= (-77.1 - 22.83)\,\mathrm{kJ\,mol^{-1}} \\
&= -99.9\,\mathrm{kJ\,mol^{-1}}
\end{aligned}
$$

(ii) $$
\begin{aligned}
(\mathrm{d}G/\mathrm{d}\xi) &= (-77.1\,\mathrm{kJ\,mol^{-1}}) + (8.314 \times 10^{-3}\,\mathrm{kJ\,K^{-1}\,mol^{-1}} \times 298.15\,\mathrm{K})\ln(10^{-5}/10^{-21}) \\
&= (-77.1 + 91.32)\,\mathrm{kJ\,mol^{-1}} \\
&= +14.2\,\mathrm{kJ\,mol^{-1}}
\end{aligned}
$$

(c) In (i), $(dG/d\xi) < 0$, so the reaction in equation 16 should have a tendency to go from left to right as written. This implies that the left-hand electrode in the cell diagram should be the anode and hence, according to the sign convention in Section 2.3.2, that the cell emf $E > 0$, as observed. In (ii) the situation is reversed, again as found experimentally.

SAQ 9 (Objectives 4, 6 and 7)

(a) For equation 52,

$$\Delta G_{\mathrm{m}}^{\ominus} = \Delta G_{\mathrm{f}}^{\ominus}(\mathrm{Ag^+, aq}) + \Delta G_{\mathrm{f}}^{\ominus}(\mathrm{Cl^-, aq}) - \Delta G_{\mathrm{f}}^{\ominus}(\mathrm{AgCl, s})$$

$$= \{77.1 + (-131.2) - (-109.8)\}\ \mathrm{kJ\ mol^{-1}}$$

$$= 55.7\ \mathrm{kJ\ mol^{-1}}$$

Using the relation $\Delta G_{\mathrm{m}}^{\ominus} = -RT \ln K^{\ominus}$,

$$\ln K_{\mathrm{sp}}^{\ominus} = -(55.7 \times 10^3\ \mathrm{J\ mol^{-1}})/(8.314\ \mathrm{J\ K^{-1}\ mol^{-1}} \times 298.15\ \mathrm{K})$$

and

$$K_{\mathrm{sp}}^{\ominus} = 1.74 \times 10^{-10}$$

(b) For AgCl, $s = c_+ = c_-$ (Section 4.7.1), so equation 58 can be rewritten as follows:

$$K_{\mathrm{sp}}^{\ominus} = \gamma_{\pm}^2 (s/c^{\ominus})^2$$

Taking the square root of both sides, this becomes

$$(K_{\mathrm{sp}}^{\ominus})^{1/2} = \gamma_{\pm}(s/c^{\ominus})$$

so

$$\gamma_{\pm} = (K_{\mathrm{sp}}^{\ominus})^{1/2}/(s/c^{\ominus})$$

whence (i) $\gamma_{\pm} = 0.891$ and (ii) $\gamma_{\pm} = 0.824$.

SAQ 10 (Objectives 8 and 9)

(a) In this case, $c(\mathrm{Ag^+}) = c(\mathrm{Cl^-}) = s = 1.48 \times 10^{-5}\ \mathrm{mol\ dm^{-3}}$; $c(\mathrm{Mg^{2+}}) = c(\mathrm{SO_4^{2-}}) = 2.00 \times 10^{-3}\ \mathrm{mol\ dm^{-3}}$, whence

$$I = \tfrac{1}{2}\{c(\mathrm{Ag^+})(+1)^2 + c(\mathrm{Cl^-})(-1)^2 + c(\mathrm{Mg^{2+}})(+2)^2 + c(\mathrm{SO_4^{2-}})(-2)^2\}/c^{\ominus}$$

$$= \{c(\mathrm{Ag^+}) + 4c(\mathrm{Mg^{2+}})\}/c^{\ominus}$$

$$= 8.014\,8 \times 10^{-3} \approx 8.01 \times 10^{-3}$$

Notice that the ionic strength is largely determined by the $MgSO_4$ in the solution.

(b) Yes, in the sense that the definition of ionic strength certainly explains how $\gamma_{\pm}$ for AgCl, and hence its solubility, can depend on the presence of other ions in the solution. Indeed, it is the variation in ionic strength (brought about by the addition of foreign ions) that effectively gives rise to the alteration of solubility shown in Figure 15.

SAQ 11 (Objective 9)

(a) If the system obeys the Debye–Hückel limiting law then equation 66 predicts that a plot of $\log(s/c^{\ominus})$ against $I^{1/2}$ (that is, y against x in Figure 20) should be a straight line, with a slope of $m = 0.51$. Further, the value of $K_{\mathrm{sp}}^{\ominus}$ can then be determined from the intercept at $I^{1/2} = 0$ – that is, the intercept c on the vertical (y) axis is given by:

$$\tfrac{1}{2}\log K_{\mathrm{sp}}^{\ominus} = \{\log(s/c^{\ominus})\} \text{ when } I^{1/2} = 0$$

The 'best' straight line through a set of experimental points should ideally be determined with the help of a least-squares analysis. You may have the facility to do this on your calculator or home computer. In this case, however, you do not have access to the raw data, so the only option is to choose the best line by eye: ours is shown dashed in Figure 32. There are two points to note:

(i) To base the extrapolation on equation 66 requires that the line through the data should have a slope of 0.51. The simplest way to ensure this is to first draw a line *with the desired slope* (through the origin, say – ours is shown in black in Figure 32), and then move your ruler parallel to this line.

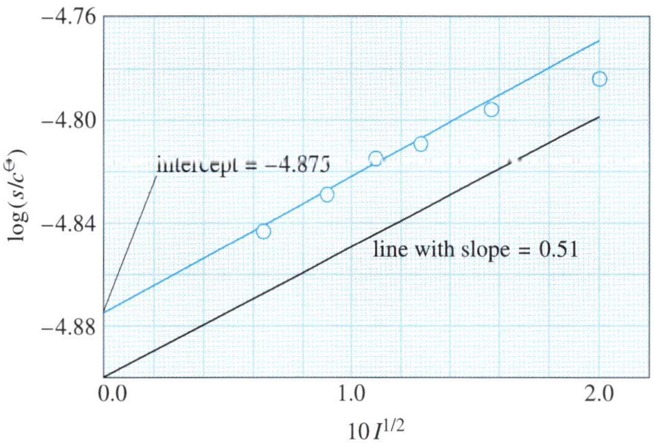

Figure 32 Determination of $K^{\ominus}_{sp}$ for AgCl.

(ii) When you do this, it becomes apparent that the final point (corresponding to the *highest* ionic strength, $I = 4.0 \times 10^{-2}$) already shows a noticeable deviation from equation 66 (and hence equation 63). We therefore ignored this point in drawing our line. From the intercept shown, $\frac{1}{2}\log K^{\ominus}_{sp} = -4.875$, so $K^{\ominus}_{sp} = 1.78 \times 10^{-10}$.

(b) Point (ii) above effectively answers this question. Clearly, the higher the ionic strength of the solution, the greater the deviations from the predictions of the Debye–Hückel limiting law.

SAQ 12 (Objectives 4, 6, 7, 8 and 9)

(a) For the reaction $BaSO_4(s) = Ba^{2+}(aq) + SO_4^{2-}(aq)$,

$$\Delta G^{\ominus}_{m} = \{-560.7 + (-744.5) - (-1\,362.1)\}\ \text{kJ mol}^{-1} = +56.9\ \text{kJ mol}^{-1}$$

Using the relation $\Delta G^{\ominus}_{m} = -RT\ln K^{\ominus}$ then gives $K^{\ominus}_{sp} = 1.074 \times 10^{-10}$.

(b) The dissociation of $BaSO_4$ is directly comparable with that of AgCl, so $K^{\ominus}_{sp}$ is related to the solubility s via equation 64:

$$K^{\ominus}_{sp} = \gamma_{\pm}^{2}(s/c^{\ominus})^2$$

Thus

$$(s/c^{\ominus})^2 = K^{\ominus}_{sp}/\gamma_{\pm}^{2}$$

or

$$s/c^{\ominus} = (K^{\ominus}_{sp})^{1/2}/\gamma_{\pm}$$

Because $BaSO_4$ is only sparingly soluble in pure water (witness the value of $K^{\ominus}_{sp}$), the saturated solution will have a very low ionic strength. Under these circumstances, the *assumption* of ideal behaviour (i.e. $\gamma_{\pm} = 1$) is a reasonable approximation, so

$$s/c^{\ominus} \approx (K^{\ominus}_{sp})^{1/2}$$

whence

$$s \approx 1.04 \times 10^{-5}\ \text{mol dm}^{-3}$$

(c) In 0.01 molar $NaNO_3$, *assume* that the solubility of $BaSO_4$ is still so small that the ionic strength of the solution is largely determined by $NaNO_3$; that is, $I = 0.01$

(since $I = c/c^{\ominus}$ for a 1 : 1 electrolyte of univalent ions). Using the Debye–Hückel limiting law (equation 63),

$$\log \gamma_{\pm} = -0.51 \times |2 \times -2| \times 0.01^{1/2} = -0.204$$

so

$$\gamma_{\pm} = 0.625$$

Substituting this value, and that for $K_{sp}^{\ominus}$ from part (a) into the expression in part (b) gives

$$s/c^{\ominus} \approx (1.074 \times 10^{-10})^{1/2}/0.625$$

whence

$$s \approx 1.66 \times 10^{-5}\ \text{mol dm}^{-3}$$

Evidently, the solubility of $BaSO_4$ is *enhanced* by the 'foreign' electrolyte – as is that of AgCl.

SAQ 13 (Objective 10)

The two values are calculated as follows:

$$q_{rev} = T\Delta S = (298.15\ \text{K}) \times (-193.2 \times 10^{-3}\ \text{J K}^{-1})$$

$$= -57.6\ \text{J}$$

$$w_{el,\ max} = \Delta H - T\Delta S = (-146.4 + 57.6)\ \text{J} = -88.8\ \text{J}$$

Yes: the value of $w_{el,\ max}$ is negative, implying the system does work on the surroundings, and thereby transfers energy to it. Equally, $\Delta G < 0$ for this change (from equation 72): evidently work can be extracted only from a spontaneous process.

SAQ 14 (Objectives 3 and 5)

The reaction in question is:

$$\tfrac{1}{2}H_2(g) + Ag^+(aq) = H^+(aq) + Ag(s) \qquad (16)$$

From equation 78, $E = -(\text{d}G/\text{d}\xi)/nF$; with $n = 1$ (from equation 16), $F = 96\,485\ \text{C mol}^{-1}$ and the values of $(\text{d}G/\text{d}\xi)$ given in the question:

(i) $$E = \frac{-(-99.9 \times 10^3\ \text{J mol}^{-1})}{(96\,485\ \text{C mol}^{-1})} = 1.04\ \text{J C}^{-1}$$

$$= 1.04\ \text{V (with the relation V} = \text{J C}^{-1})$$

(ii) Similarly, in this case $E = -0.15$ V.

This confirms that the conditions $(\text{d}G/\text{d}\xi) < 0$ and $E > 0$ are equivalent criteria for a spontaneous change in the system, *as written* in equation 16. If $(\text{d}G/\text{d}\xi) > 0$ or $E < 0$, then the reverse process occurs. The approximation is contained in the values of $(\text{d}G/\text{d}\xi)$ taken from SAQ 8: both were calculated on the assumption that the electrolytes behave ideally; that is, $\gamma_{\pm} = 1$.

SAQ 15 (Objective 11)

(a) From the definition in the text, $E^{\ominus}$ for the $(Zn^{2+}|Zn)$ couple refers to the reaction:

$$Zn^{2+}(aq) + H_2(g) = Zn(s) + 2H^+(aq)$$

Since $\Delta G_f^{\ominus}$ is zero for $H_2(g)$ and Zn(s), and is defined to be zero for $H^+(aq)$,

$$\Delta G_m^{\ominus} = -\Delta G_f^{\ominus}(Zn^{2+}, aq) = +147.1\ \text{kJ mol}^{-1}$$ (from the S342 *Data Book*)

Using the relation $\Delta G_m^{\ominus} = -nFE^{\ominus}$, with $n = 2$,

$$E^{\ominus}(Zn^{2+}|Zn) = -\{(147.1 \times 10^3)/(2 \times 96\,485)\}\ \text{V} = -0.76\ \text{V}$$

The negative value of $E^\ominus$ implies that the spontaneous process in this system is the reverse of the reaction above. In other words, under standard conditions at 298.15 K, it is thermodynamically favourable for H^+(aq) ions to oxidize zinc to Zn^{2+}(aq).

(b) No. Dividing through the stoichiometry by two has the effect of halving *both* the value of $\Delta G_m^\ominus$ *and* that of n: the value of $E^\ominus$ is unchanged. Thus, although $\Delta G_m^\ominus$ does depend on the stoichiometry (that is, the coefficient of e), $E^\ominus$ does not, because n appears in the equation (equation 80) that relates these two quantities.

SAQ 16 (Objectives 12 and 14)

From the closing paragraphs of Section 6.1.2, reduction is thermodynamically favourable if the $E^\ominus$ value of the couple containing the reduced state is more negative than the $E^\ominus$ value of the couple containing the oxidized state. This gives:
(a) favourable; (b) unfavourable; (c) favourable.

As indicated in the question, reduction (c) does not actually occur, despite the prediction above. This emphasizes the fact that $E^\ominus$ values are subject to the *same* limitations as the $\Delta G_m^\ominus$ values from which they are often derived: both are thermodynamic quantities, and predictions made using them can be nullified by kinetic effects. (More on this in Section 8.)

SAQ 17 (Objective 12)

This question is based on the same analysis as that in SAQ 16, save that the couples involved are more complex. The answers are: (a) unfavourable; (b) favourable; (c) favourable; (d) favourable.

(a) Following the procedure in the text, the reaction given can be 'decomposed' into two reduction processes, as follows:

RHE: $H^+(aq) + e = \frac{1}{2}H_2(g)$

LHE: $\frac{1}{2}Cl_2(g) + e = Cl^-(aq)$

RHE – LHE: $H^+(aq) + Cl^-(aq) = \frac{1}{2}H_2(g) + \frac{1}{2}Cl_2(g)$

So $E^\ominus(\text{overall}) = E^\ominus_{RHE} - E^\ominus_{LHE} = -1.36\ \text{V}$.

The same result could be arrived at more simply by recognizing that the reaction given is just the *reverse* of the ($Cl_2|Cl^-$) couple, since 'e' is equivalent to $\{\frac{1}{2}H_2(g) - H^+(aq)\}$.

(b) Here, the reaction given is equivalent to the difference RHE – LHE, with

RHE: $2Fe^{3+}(aq) + 2e = 2Fe^{2+}(aq)$

LHE: $Sn^{4+}(aq) + 2e = Sn^{2+}(aq)$

RHE – LHE : $2Fe^{3+}(aq) + Sn^{2+}(aq) = 2Fe^{2+}(aq) + Sn^{4+}(aq)$

So $E^\ominus(\text{overall}) = E^\ominus_{RHE} - E^\ominus_{LHE} = \{0.77 - (0.15)\}\ \text{V} = 0.62\ \text{V}$.

Notice, again, that the *stoichiometry* of the overall reaction is established by combining the couples so as to eliminate 'e', but the corresponding values of $E^\ominus$ are simply subtracted directly.

(c) This time, the reaction is equivalent to RHE – LHE, with

RHE: $Cr_2O_7^{2-}(aq) + 14H^+(aq) + 6e = 2Cr^{3+}(aq) + 7H_2O(l)$

LHE: $6Fe^{3+}(aq) + 6e = 6Fe^{2+}(aq)$

So $E^\ominus(\text{overall}) = E^\ominus_{RHE} - E^\ominus_{LHE} = \{1.36 - (0.77)\}\ \text{V} = 0.59\ \text{V}$.

Incidentally, this reaction is used in practice to estimate the iron content in ores.

(d) The couples involved here are fairly obvious from the selection available in Table 6.

RHE: $PbO_2(s) + SO_4^{2-}(aq) + 4H^+(aq) + 2e = PbSO_4(s) + 2H_2O(l)$

LHE: $PbSO_4(s) + 2e = Pb(s) + SO_4^{2-}(aq)$

So $E^{\ominus}(\text{overall}) = E^{\ominus}_{RHE} - E^{\ominus}_{LHE} = \{1.69 - (-0.36)\}\ \text{V} = 2.05\ \text{V}.$

In this case the two half-reactions have immense practical importance: they take place at the cathode and anode, respectively, of the lead/acid battery in a car.

SAQ 18 (Objective 12)

This is one example of an important class of reactions representing the 'solubility' of a sparingly soluble compound in water. In each case, the overall reaction can be decomposed into a metal/insoluble salt couple and the corresponding metal/aqueous metal cation couple, as follows:

RHE: $AgCl(s) + e = Ag(s) + Cl^-(aq)$

LHE: $Ag^+(aq) + e = Ag(s)$

RHE – LHE: $AgCl(s) = Ag^+(aq) + Cl^-(aq)$ (52)

So $E^{\ominus}(\text{overall}) = E^{\ominus}_{RHE} - E^{\ominus}_{LHE} = \{0.22 - (0.80)\}\ \text{V} = -0.58\ \text{V} = -0.58\ \text{J C}^{-1}$

Using equation 87, with $n = 1$,

$$\ln K^{\ominus} = nFE^{\ominus}/RT = \frac{1 \times (96\,485\ \text{C mol}^{-1}) \times (-0.58\ \text{J C}^{-1})}{(8.314\ \text{J K}^{-1}\ \text{mol}^{-1}) \times (298.15\ \text{K})} = -22.576$$

$$K^{\ominus} = K^{\ominus}_{sp} = 1.57 \times 10^{-10}$$

[*Note* The discrepancy between this value and that calculated in SAQ 9, 1.74×10^{-10}, arises from our use of $E^{\ominus}$ values 'rounded' to two decimal places.]

SAQ 19 (Objectives 4, 13 and 14)

(a) If iron corrodes by being oxidized to the $Fe^{2+}(aq)$ ion, then the corrosion reaction can be written:

RHE: $\frac{1}{2}O_2(g) + 2H^+(aq) + 2e = H_2O(l)$ (91)

LHE: $Fe^{2+}(aq) + 2e = Fe(s)$

RHE – LHE: $\frac{1}{2}O_2(g) + 2H^+(aq) + Fe(s) = Fe^{2+}(aq) + H_2O(l)$

Iron will have a thermodynamic tendency to corrode if $E_{RHE} - E_{LHE} > 0$ – that is, if $E(91)$ is greater (more positive) than $E(Fe^{2+}|Fe)$ under the stated conditions.

At pH = 5,

$$E(91) = (1.23 - (0.059\,2 \times 5))\ \text{V (from equation 94)}$$

$$= 0.93\ \text{V}$$

The required value of $E(Fe^{2+}|Fe)$ is calculated by applying the Nernst equation to the $(Fe^{2+}|Fe)$ couple with $a(Fe^{2+}) \approx c(Fe^{2+})/c^{\ominus} = 10^{-6}$. This gives:

$$E(Fe^{2+}|Fe) = E^{\ominus}(Fe^{2+}|Fe) - \frac{RT}{2F}\ln\frac{a(Fe)}{a(Fe^{2+})}$$

$$= E^{\ominus}(Fe^{2+}|Fe) + (RT/2F)\ln a(Fe^{2+}),\ \text{since } a(Fe) = 1$$

$$= (-0.46 + 0.0128 \ln 10^{-6})\ \text{V}$$

$$= -0.64\ \text{V}$$

Similar analyses give: (b) $E(Al^{3+}|Al) = -1.80$ V (with $n = 3$ in the Nernst expression); (c) $E(Sn^{2+}|Sn) = -0.32$ V.

In each case, $E(91)$ *is* more positive than the electrode potential of the metallic couple. In other words, this simple analysis predicts that all three metals have a tendency to corrode under these conditions. This accords with the observed behaviour of iron, but not with that of aluminium (saucepans, etc.) or tin (coatings of which are used to *protect* iron in tin cans, etc.). These results again underline the limitations of a purely thermodynamic approach to this extremely important problem. (See also Section 8.)

SAQ 20 (Objectives 4, 13 and 14)

The short answer is 'Yes – but with some important reservations'. Given the discussion in the text, it should be clear that the values of $E^{\ominus}$ for both couples can be related to the emf of the *cell* discussed in Section 3, that is

$$\text{Pt, } H_2(g)|H^+(aq)|Ag^+(aq)|Ag(s)$$

However, $E^{\ominus}$ values require a close specification of the composition of the cell. For a start, the hydrogen electrode has to be under standard conditions (with $a(H^+) = 1$ and $p(H_2) = 1$ bar). For the cell emf to equate with $E^{\ominus}(Ag^+|Ag)$, the S.H.E. would have to be combined with the set-up depicted in Figure 28a (with $a(Ag^+) = 1$). Similar reasoning indicates that the S.H.E. combined with a half-cell like the one depicted in Figure 28c (with $a(S^{2-}) = 1$, in the presence of solid Ag_2S) would – in principle, at least – give a cell emf that equates with $E^{\ominus}$ for the silver/silver sulfide couple in equation 96.

Reference to the experimental observations in Table 2 (Section 3) reveals that the conditions specified there do *not* fulfill these requirements. Nevertheless, the video sequence demonstrated that the polarity of the cell above could be *reversed* by precipitating out Ag_2S – an observation that *is* consistent with the change of *sign*, from $E^{\ominus}(Ag^+|Ag) = +0.80$ V to $E^{\ominus} = -0.66$ V for the couple in equation 96.

SAQ 21 (Objectives 2, 12, 13 and 14)

(a) False; (b) true; (c) true.

(a) For the statement to be true, electrons would have to be generated at the LHE (i.e. this should be the anode at which oxidation occurs) and consumed at the RHE. In other words, the implied cell reaction:

$$3Ni(s) + 2Al^{3+}(aq) = 3Ni^{2+}(aq) + 2Al(s)$$

should be a spontaneous process. But $E^{\ominus}(\text{cell}) = E^{\ominus}_{RHE} - E^{\ominus}_{LHE} = E^{\ominus}(Al^{3+}|Al) - E^{\ominus}(Ni^{2+}|Ni)$. Consulting Table 5, it turns out that $E^{\ominus}(Al^{3+}|Al)$ is *more negative* than $E^{\ominus}(Ni^{2+}|Ni)$, so $E^{\ominus}(\text{cell}) < 0$, implying that the *reverse* process is the spontaneous reaction. Thus the statement is false.

(b) The couples implicit in this statement can be spotted from the selection available in Table 6:

$$MnO_4^-(aq) + 8H^+(aq) + 5e = Mn^{2+}(aq) + 4H_2O(l);\ E^{\ominus} = 1.51 \text{ V}$$

$$Sn^{4+}(aq) + 2e = Sn^{2+}(aq);\ E^{\ominus} = 0.15 \text{ V}$$

Since $E^{\ominus}$ for the permanganate couple is much more positive than $E^{\ominus}(Sn^{4+}|Sn^{2+})$, point 4 in Box 6 suggests that the statement can be taken as true, even though the precise conditions (pH, etc.) are not specified – see the discussion in Section 6.2.1. (Again, think of the couples above as RHE and LHE, respectively, if this helps.)

(c) This statement involves the following couples:

$$\tfrac{1}{2}O_2(g) + 2H^+(aq) + 2e = H_2O(l);\ E = (1.23 - 0.0592\text{ pH}) \text{ V} = 1.17 \text{ V at pH} = 1 \qquad (94)$$

$$Fe^{3+}(aq) + e = Fe^{2+}(aq);\ E^{\ominus} = 0.77 \text{ V}$$

Although the precise composition of the solution is not specified, these electrode potentials suggest that substantial amounts of $Fe^{2+}(aq)$ should not be capable of sustained existence in the presence of atmospheric oxygen. In practice, the oxidation does occur, but it is slow. This is why stock solutions of $Fe^{2+}(aq)$ must be re-standardized before they are used for titrations.

SAQ 22 (Objectives 2, 4, 12, 13 and 14)

(a) Here, as in SAQ 21a, it is straightforward to 'read' the cell diagram from left to right, and hence write down the implied cell reaction directly, as:

$$2Ag(s) + Ni^{2+}(aq) = 2Ag^+(aq) + Ni(s)$$

Exactly the same result is obtained by following the alternative procedure emphasized throughout Section 6, as:

RHE:	$Ni^{2+}(aq) + 2e = Ni(s)$;	$E^\ominus = -0.24$ V
LHE:	$2Ag^+(aq) + 2e = 2Ag(s)$;	$E^\ominus = +0.80$ V

RHE – LHE: $Ni^{2+}(aq) - 2Ag^+(aq) = Ni(s) - 2Ag(s)$

which can be rearranged to give the implied cell reaction above.

(b) Using the $E^\ominus$ values above (taken from Table 5),

$$E^\ominus(\text{cell}) = E^\ominus_{\text{RHE}} - E^\ominus_{\text{LHE}}$$

$$= \{-0.24 - (0.80)\}\text{ V} = -1.04\text{ V}$$

But the cell is *not* set up under standard conditions. Applying the Nernst equation *to the implied cell reaction* gives:

$$E = E^\ominus - \frac{RT}{2F}\ln\left(\frac{\{a(Ag^+)\}^2\, a(Ni)}{\{a(Ag)\}^2\, a(Ni^{2+})}\right)$$

For the solids, Ag and Ni, $a = 1$ by definition. But strictly, $a(Ag^+) = \gamma_\pm c(Ag^+)/c^\ominus$ and $a(Ni^{2+}) = \gamma_\pm c(Ni^{2+})/c^\ominus$, where $\gamma_\pm$ represents the mean ionic activity coefficient of the electrolyte in each half-cell. Because you are not given any information about these quantities, further progress requires the *assumption of ideal behaviour* – that is, $\gamma_\pm = 1$. Then,

$$a(Ag^+) = c(Ag^+)/c^\ominus = 0.1 \text{ (with } c^\ominus = 1\text{ mol dm}^{-3})$$

$$a(Ni^{2+}) = c(Ni^{2+})/c^\ominus = 1.0$$

Substituting back into the Nernst equation gives:

$$E = -1.04\text{ V} - (0.0128\text{ V})\ln\left(\frac{(0.1)^2 \times 1}{1^2 \times 1.0}\right)$$

$$= (-1.04 + 0.06)\text{ V}$$

$$= -0.98\text{ V}$$

The negative sign indicates that the spontaneous process is the *reverse* of the implied cell reaction, as:

$$2Ag^+(aq) + Ni(s) = 2Ag(s) + Ni^{2+}(aq)$$

SAQ 23 (Objectives 4 and 13)

Applying the Nernst equation to the $(H^+|H_2)$ couple:

$$H^+(aq) + e = \tfrac{1}{2}H_2(g)$$

$$E = E^\ominus - \frac{RT}{F}\ln\left\{\frac{a(H_2)^{1/2}}{a(H^+)}\right\}$$

Since $a(H_2) = p(H_2)/p^\ominus = 1$ (with the assumption in the question) this becomes (cf. the discussion of the 'oxygen' couple in Section 6.2.3):

$$E = E^\ominus + (RT/F)\ln a(H^+)$$

$$= E^\ominus + (0.0592\text{ V})\log a(H^+)$$

$$= E^\ominus - (0.0592\text{ V})\,\text{pH}$$

or

$$E/\text{V} = -0.0592\,\text{pH} \text{ (since } E^\ominus = 0)$$

SAQ 24 (Objectives 4, 6, 12 and 13)

Following the analysis in Section 6.2.3, $E^\ominus$ for the couple in the question can be considered to be the value of E for the $(Zn^{2+}|Zn)$ couple when the activity of *hydroxide ions* is unity ($a(OH^-) = 1$), such that $a(Zn^{2+})$ is determined by the solubility product of $Zn(OH)_2$, as follows:

$$Zn(OH)_2(s) = Zn^{2+}(aq) + 2OH^-(aq)$$

$$K_{sp}^\ominus = a(Zn^{2+}) \times \{a(OH^-)\}^2 = 7.61 \times 10^{-17}, \text{ since } a\{Zn(OH)_2\} = 1$$

So

$$a(Zn^{2+}) = 7.61 \times 10^{-17}, \text{ with } a(OH^-) = 1$$

Application of the Nernst equation to $(Zn^{2+}|Zn)$ then gives:

$$Zn^{2+}(aq) + 2e = Zn(s)$$

$$E = E^\ominus - (RT/2F)\ln\{a(Zn)/a(Zn^{2+})\}$$

$$= E^\ominus + (RT/2F)\ln a(Zn^{2+}), \text{ since } a(Zn) = 1$$

$$= \{-0.76 + (-0.48)\}\text{ V} = -1.24\text{ V}$$

Check that this is the value of $E^\ominus$ for the zinc hydroxide/zinc couple listed in your S342 *Data Book*.

SAQ 25 (Objectives 2, 10 and 12)

The spontaneous cell reaction must be as follows:

RHE:	$HgO(s) + H_2O(l) + 2e = Hg(l) + 2OH^-(aq)$	$E^\ominus = +0.098$ V
LHE:	$Zn(OH)_2(s) + 2e = Zn(s) + 2OH^-(aq)$	$E^\ominus = -1.24$ V
RHE – LHE:	$HgO(s) + H_2O(l) + Zn(s) = Hg(l) + Zn(OH)_2(s)$	$E^\ominus = 1.34$ V

Thus oxidation takes place at the *zinc anode*, and reduction at the *mercury cathode*. From equations 72 and 80,

$$w_{el,\,max} = \Delta G_m^\ominus = -nFE^\ominus$$

$$= -2 \times 96\,485 \times 1.34 \times 10^{-3}\text{ kJ mol}^{-1}$$

$$= -258.6\text{ kJ mol}^{-1}$$

Neither $H^+(aq)$ nor $OH^-(aq)$ ions appear in the overall cell reaction, so the cell emf should be independent of pH. In commercial cells, the electrolyte is actually concentrated potassium hydroxide, and the cell potential is close to that calculated here.

SAQ 26 (Objectives 11 and 15)

There is an important general point here: when using $E^\ominus$ values in conjunction with other types of thermodynamic data it is always important – *and sometimes crucial* – to remember that 'e' is shorthand for $\{\frac{1}{2}H_2(g) - H^+(aq)\}$. Thus, $E^\ominus(Cl^-|AgCl|Ag)$ refers to the reaction:

$$AgCl(s) + \tfrac{1}{2}H_2(g) = Ag(s) + Cl^-(aq) + H^+(aq) \qquad (97)$$

Using the relation $\Delta G_m^\ominus = -nFE^\ominus$ (equation 80), with $n = 1$ and $E^\ominus = 0.222\,3$ V at 298.15 K,

$$\Delta G_m^\ominus(97) = -21.4\text{ kJ mol}^{-1}$$

But, for reaction 97,

$$\Delta G_m^\ominus(97) = \Delta G_f^\ominus(Ag, s) + \Delta G_f^\ominus(Cl^-, aq) + \Delta G_f^\ominus(H^+, aq) - \Delta G_f^\ominus(AgCl, s) - \tfrac{1}{2}\Delta G_f^\ominus(H_2, g)$$

Since $\Delta G_f^\ominus = 0$ for the elements and for $H^+(aq)$, this can be reorganized to give

$$\Delta G_f^\ominus(Cl^-, aq) = \Delta G_m^\ominus(97) + \Delta G_f^\ominus(AgCl, s)$$

$$= \{-21.4 + (-109.8)\}\text{ kJ mol}^{-1}$$

$$= -131.2\text{ kJ mol}^{-1} \text{ (cf. the value in the S342 } \textit{Data Book}\text{)}$$

SAQ 27 (Objective 15)

Equation 99 can be rearranged as follows, with $E^\ominus = 0.222\,3$ V and $E = 468.6$ mV = 0.468 6 V when $c = 9.14 \times 10^{-3}$ mol dm^{-3}:

$$2(RT/F)\ln\gamma_\pm = E^\ominus - E - 2(RT/F)\ln(c/c^\ominus)$$

or

$$\ln\gamma_\pm = (F/2RT)(E^\ominus - E) - \ln(c/c^\ominus)$$

$$= 19.462 \times (-0.246\,3) - (-4.695\,1) = -0.098\,4$$

$$\gamma_\pm = 0.906$$

This again demonstrates the importance of this type of cell, with a single electrolyte. Once $E^\ominus$ for such a cell is known, then the value of $\gamma_\pm$ for a particular concentration of the electrolyte is readily determined by measuring the corresponding emf of the cell.

SAQ 28 (Objectives 15 and 16)

The value of $\Delta S_m^\ominus$ follows directly from equation 102, with $(T_2 - T_1) = 10$ K, and $(E_2^\ominus - E_1^\ominus) = (0.219\,1 - 0.225\,6)$ V = $-0.006\,5$ V = $-0.006\,5$ J C^{-1}, so (with $n = 1$):

$$\Delta S_m^\ominus = \frac{(-0.006\,5\text{ J C}^{-1})\times(1\times 96\,485\text{ C mol}^{-1})}{10\text{ K}}$$

$$= -62.7\text{ J K}^{-1}\text{ mol}^{-1}$$

The warning in the question relates back to the point made in the answer to SAQ 26 – that is, $E^\ominus(Cl^-|AgCl|Ag)$ relates to the reaction in equation 97, so:

$$\Delta S_m^\ominus = S^\ominus(\text{Ag, s}) + S^\ominus(\text{Cl}^-, \text{aq}) + S^\ominus(\text{H}^+, \text{aq}) - S^\ominus(\text{AgCl, s}) - \tfrac{1}{2}S^\ominus(\text{H}_2, \text{g})$$

So

$$S^\ominus(\text{Cl}^-, \text{aq}) = \Delta S_m^\ominus - S^\ominus(\text{Ag, s}) - S^\ominus(\text{H}^+, \text{aq}) + S^\ominus(\text{AgCl, s}) + \tfrac{1}{2}S^\ominus(\text{H}_2, \text{g})$$

$$= (-62.7 - 42.6 - 0 + 96.2 + \tfrac{1}{2} \times 130.7)\text{ J K}^{-1}\text{ mol}^{-1}$$

$$= 56.3\text{ J K}^{-1}\text{ mol}^{-1}$$

Assuming $\Delta S_m^\ominus$ *and* $\Delta H_m^\ominus$ *do not change with temperature*, $\Delta H_m^\ominus$ (at 298.15 K) can then be calculated from $\Delta G_m^\ominus$ at either 20 °C (293.15 K) or 30 °C (303.15 K): for example,

$$\Delta G_m^\ominus(293.15\text{ K}) = -21.8\text{ kJ mol}^{-1}\ (\text{from } E^\ominus = 0.225\,6\text{ V})$$

So

$$\Delta H_m^\ominus = \Delta G_m^\ominus + T\Delta S_m^\ominus$$

$$= (-21.8\text{ kJ mol}^{-1}) + (293.15\text{ K}) \times (-62.7 \times 10^{-3}\text{ kJ K}^{-1}\text{ mol}^{-1})$$

$$= -40.2\text{ kJ mol}^{-1}$$

SAQ 29 (Objectives 14 and 16)

For the reaction in equation 89,

$$\Delta S_m^\ominus = 2S^\ominus(\text{PbSO}_4, \text{s}) + 2S^\ominus(\text{H}_2\text{O, l}) - S^\ominus(\text{Pb, s}) - S^\ominus(\text{PbO}_2, \text{s}) - 4S^\ominus(\text{H}^+, \text{aq}) - 2S^\ominus(\text{SO}_4^{2-}, \text{aq})$$

$$= \{2(148.6) + 2(69.9) - (64.8) - (68.3) - 4(0) - 2(20.1)\}\text{ J K}^{-1}\text{ mol}^{-1}$$

$$= 263.7\text{ J K}^{-1}\text{ mol}^{-1}$$

Then, using equation 102, with $T_1 = 298.15$ K, $T_2 = 273.15$ K (0 °C), $n = 2$ (from equation 89), and $E_1^\ominus = 2.05$ V (given),

$$E_2^\ominus(0\,°\text{C}) = \frac{(263.7\text{ J K}^{-1}\text{ mol}^{-1})\times(-25\text{ K})}{(2\times 96\,485\text{ C mol}^{-1})} + 2.05\text{ V}$$

$$= (-0.034 + 2.05)\text{ V} = 2.02\text{ V}$$

Evidently the emf falls as the temperature is lowered ($\Delta S_m^\ominus > 0$), but this effect alone is not sufficient to explain the reduced performance of a car battery at 0 °C (and below). A more important factor is the effect of temperature on the *kinetics* of the cell reaction, and hence the *rate* at which energy is converted. This point is taken up in a different context in Section 8, and explored further in Block 8.

SAQ 30 (Objectives 4, 13 and 14)

This question is best approached by determining the electrode potentials of the relevant couples under the stated conditions (assuming $T = 298.15$ K, throughout):

(i) $\frac{1}{2}O_2(g) + 2H^+(aq) + 2e = H_2O(l)$

From equation 94, $E = (1.23 - 0.059\,2\,\text{pH})$ V so, at pH = 3,

$$E = 1.05 \text{ V}$$

(ii) $Zn^{2+}(aq) + 2e = Zn(s)$

Applying the Nernst equation to this couple gives (see the answer to SAQ 24):

$$E = E^\ominus + (RT/2F) \ln a(Zn^{2+})$$

Strictly, $a(Zn^{2+}) = \gamma_\pm c(Zn^{2+})/c^\ominus$. *Assuming ideal behaviour* (i.e. $\gamma_\pm = 1$, for want of any information about the activity coefficient), $a(Zn^{2+}) \approx c(Zn^{2+})/c^\ominus = 2$ (with $c^\ominus = 1 \text{ mol dm}^{-3}$). With $E^\ominus = -0.76$ V, therefore,

$$E = (-0.76 + 0.009) \text{ V} = -0.75 \text{ V}$$

(iii) $H^+(aq) + e = \frac{1}{2}H_2(g)$

From the expression in SAQ 23, $E = (-0.059\,2\,\text{pH})$ V so, at pH = 3,

$$E = -0.18 \text{ V}$$

Following the discussion in the text, the minimum potentials required are given by $E = E_{RHE} - E_{LHE}$, where the LHE is always the oxygen couple in (i), but the RHE can be either (a) ($Zn^{2+}|Zn$), or (b) ($H^+|H_2$). Thus:

(a) $E = E_{RHE} - E_{LHE}$
$= (-0.75 - 1.05) \text{ V} = -1.8 \text{ V}$

(b) $E = E_{RHE} - E_{LHE}$
$= (-0.18 - 1.05) \text{ V} = -1.23 \text{ V}$

which confirms that the minimum potential for electrolysis of the solvent (water) is independent of pH – as stated in the text.

Thus according to thermodynamics alone, under these conditions a potential greater than 1.8 V must be applied across the cell in order to deposit zinc metal at the cathode (and liberate oxygen at the anode). However, this potential is also more than enough to liberate hydrogen gas at the cathode (that is, it is greater than 1.23 V). So these data suggest it is impossible to obtain zinc metal *alone* by this type of electrolysis.

ANSWERS TO EXERCISES

Exercise 1 (revision)

$$N_2O_4(g) = 2NO_2(g) \qquad (19)$$

Part 1 (a) Assuming $\Delta H_m^\ominus$ and $\Delta S_m^\ominus$ for reaction 19 do not change with temperature, and using $\Delta G_m^\ominus(T) = \Delta H_m^\ominus - T\Delta S_m^\ominus$,

$$\Delta G_m^\ominus(325\text{ K}) = (57.2\text{ kJ mol}^{-1}) - (325\text{ K}) \times (175.9 \times 10^{-3}\text{ kJ K}^{-1}\text{ mol}^{-1})$$

$$= +0.032\,5\text{ kJ mol}^{-1}$$

So

$$\ln K^\ominus = -\Delta G_m^\ominus / RT = -0.012\,0$$

and

$$K^\ominus = 0.988$$

Notice that at this temperature $\Delta G_m^\ominus \approx 0\text{ kJ mol}^{-1}$, so $K^\ominus \approx 1$.

For reaction 19: $K_p = \{p(NO_2)\}^2/p(N_2O_4)$

In Block 1, $K^\ominus$ was identified with the *dimensionless* ratio obtained by dividing each partial pressure in the expression for K_p by a standard pressure, $p^\ominus = 1$ bar. In this case,

$$K^\ominus = \frac{\{p(NO_2)/p^\ominus\}^2}{p(N_2O_4)/p^\ominus} = \frac{\{p(NO_2)\}^2}{p(N_2O_4)} \times \frac{1}{p^\ominus} = K_p/p^\ominus$$

So

$$K_p = K^\ominus \times p^\ominus = 0.988\text{ bar}$$

We said in Block 1 that this step implicitly assumes that the gases involved behave *ideally*: this important assumption is taken up in Section 4.6.

(b) Define $y = p(NO_2)/p_{tot}$ at equilibrium; then

$$p(NO_2) = yp_{tot}, \text{ and } p(N_2O_4) = p_{tot} - p(NO_2) = p_{tot}(1 - y)$$

So

$$K_p = \frac{(yp_{tot})^2}{(1-y)p_{tot}} = \frac{y^2 p_{tot}}{(1-y)}$$

With $y = 0.616$ and $p_{tot} = p^\ominus = 1$ bar, this yields

$$K_p = \frac{(0.616)^2}{(1-0.616)} \times (1\text{ bar}) = 0.988\text{ bar} \quad \text{(as in (a) above)}$$

From the definition of partial pressure (for an ideal gas) given in Block 1:

$$y = p(NO_2)/p_{tot} = n(NO_2)/n_{tot}$$

So y is the amount of NO_2 in the equilibrium mixture as a *fraction* of the total amount of the different gases then present.

Part 2 (a) From equation 19, $\nu(N_2O_4) = -1$, $\nu(NO_2) = +2$: products have positive coefficients, reactants have negative ones.

(b) $\xi = (n_Y - n_{Y,0})/\nu_Y$

Suppose Y is N_2O_4, then

Initially

$$n(N_2O_4) = n_0(N_2O_4) = 1\ \text{mol, so}\ \xi = 0$$

Finally (if *all* the N_2O_4 is converted)

$$n(N_2O_4) = 0$$

so

$$\xi = \{n(N_2O_4) - n_0(N_2O_4)\}/\nu(N_2O_4) = (0 - 1\ \text{mol})/-1 = +1\ \text{mol}.$$

In other words, ξ runs from 0 to 1 mol.

Similarly, shifting attention to NO_2, then

Initially

$$n(NO_2) = n_0(NO_2) = 0,\ \text{so}\ \xi = 0$$

Finally (if 1 mol of N_2O_4 is completely converted into 2 mol of NO_2)

$$n(NO_2) = 2\ \text{mol}$$

so

$$\xi = \{n(NO_2) - n_0(NO_2)\}/\nu(NO_2) = (2\ \text{mol} - 0)/2 = +1\ \text{mol}$$

This result reinforces the value of the quantity 'extent of reaction', ξ: it is a measure of how far a reaction has gone that is the *same* for all substances participating in the reaction – whether reactant *or* product.

(c) Rearranging the definition of ξ gives $n_Y = n_{Y,0} + \nu_Y\xi$

Substance	ν_Y	$n_{Y,0}$	n_Y(at equilibrium)
N_2O_4	−1	1 mol	$1\ \text{mol} - \xi_e = \{n(N_2O_4)\}_e$
NO_2	+2	0	$0 + 2\xi_e = \{n(NO_2)\}_e$
		total	$1\ \text{mol} + \xi_e = (n_{tot})_e$

From Part 1(b):

$$\text{equilibrium yield}\ y = \frac{\{n(NO_2)\}_e}{(n_{tot})_e} = \frac{2\xi_e}{(1\ \text{mol} + \xi_e)}$$

so

$$\frac{2\xi_e}{(1\ \text{mol} + \xi_e)} = 0.616$$

and

$$\xi_e = (0.616\ \text{mol})/(2 - 0.616) = 0.445\ \text{mol}$$

Exercise 2 (Objectives 4, 5 and 6)

(a) For the reaction in equation 67,

$$K^{\ominus} = \left\{\frac{a(\mathrm{Zn}^{2+})a(\mathrm{Cu})}{a(\mathrm{Cu}^{2+})a(\mathrm{Zn})}\right\}_{\mathrm{e}} = \left\{\frac{a(\mathrm{Zn}^{2+})}{a(\mathrm{Cu}^{2+})}\right\}_{\mathrm{e}}, \text{ since } a(\mathrm{Cu}) = a(\mathrm{Zn}) = 1$$

Thus $Q = a(\mathrm{Zn}^{2+})/a(\mathrm{Cu}^{2+})$ and equation 51 becomes

$$(\mathrm{d}G/\mathrm{d}\xi) = \Delta G_{\mathrm{m}}^{\ominus} + RT\ln\{a(\mathrm{Zn}^{2+})/a(\mathrm{Cu}^{2+})\} \quad (103)$$

Notice that this expression raises a problem, in that it contains the activities of *individual* aqueous ions – quantities that we claimed (in Section 4.7.2) have no meaning. As indicated there, the problem is handled by introducing the *mean* ionic activity coefficient of the electrolyte in question. Thus, for example, $a(\mathrm{Zn}^{2+})$ is taken to be equal to $\gamma_{\pm}\{c(\mathrm{Zn}^{2+})/c^{\ominus}\}$, and so on.

(b) From the expression above, if $a(\mathrm{Zn}^{2+}) = a(\mathrm{Cu}^{2+}) = 1$, then $Q = 1$, $RT\ln Q = 0$ and $(\mathrm{d}G/\mathrm{d}\xi) = \Delta G_{\mathrm{m}}^{\ominus} = -212.6\ \mathrm{kJ\ mol^{-1}}$; that is, $\Delta G_{\mathrm{m}}^{\ominus}$ is the value of $(\mathrm{d}G/\mathrm{d}\xi)$ 'under standard conditions'.

(c) Taking the hint from part (a) above:

$$a(\mathrm{Zn}^{2+}) = \gamma_{\pm}\{c(\mathrm{Zn}^{2+})/c^{\ominus}\} = 0.148 \times 0.1 = 0.014\,8$$

$$a(\mathrm{Cu}^{2+}) = \gamma_{\pm}\{c(\mathrm{Cu}^{2+})/c^{\ominus}\} = 0.043 \times 1.0 = 0.043$$

So, from equation 103,

$$(\mathrm{d}G/\mathrm{d}\xi) = \{-212.6 + (8.314 \times 298.15 \times 10^{-3})\ln(0.0148/0.043)\}\ \mathrm{kJ\ mol^{-1}}$$

$$= (-212.6 - 2.6)\ \mathrm{kJ\ mol^{-1}}$$

$$= -215.2\ \mathrm{kJ\ mol^{-1}}$$

(d) From equation 17,

$$\ln K^{\ominus} = -\Delta G_{\mathrm{m}}^{\ominus}/RT = \frac{-(-212.6 \times 10^{3})\ \mathrm{J\ mol^{-1}}}{(8.314\ \mathrm{J\ K^{-1}\ mol^{-1}} \times 298.15\ \mathrm{K})} = 85.767$$

so

$$K^{\ominus} = \{a(\mathrm{Zn}^{2+})/a(\mathrm{Cu}^{2+})\}_{\mathrm{e}} = 1.77 \times 10^{37}$$

The spontaneous cell reaction will be reversed if $(\mathrm{d}G/\mathrm{d}\xi)$, as written in part (a), becomes positive; that is, if the ratio of activities actually present exceeds the equilibrium value (or $Q > K^{\ominus}$, see Figure 14 in Section 4.6):

$$a(\mathrm{Zn}^{2+})/a(\mathrm{Cu}^{2+}) > 1.77 \times 10^{37}$$

or

$$a(\mathrm{Cu}^{2+}) < 6 \times 10^{-38} a(\mathrm{Zn}^{2+})$$

For any realistic value of $a(\mathrm{Zn}^{2+})$, this represents a *minute* concentration of $\mathrm{Cu}^{2+}(\mathrm{aq})$ – one that is effectively impossible to achieve in practice. Thus, the spontaneous cell reaction *cannot* be reversed by adjusting the concentrations within the cell, unlike the example in SAQ 8.

(e) The answers to parts (a), (b) and (d) suggest two points:

(i) $\Delta G_{\mathrm{m}}^{\ominus}$ is the value of $(\mathrm{d}G/\mathrm{d}\xi)$ when all reactants and products are in their standard states (i.e. at unit activity): in other words, the criterion $\Delta G_{\mathrm{m}}^{\ominus} < 0$ applies when the reaction occurs 'under standard conditions';

(ii) Under other circumstances, the value of $\Delta G_{\mathrm{m}}^{\ominus}$ will dominate the expression on the right-hand side of equation 51 provided it is sufficiently large and negative – as here (or large and positive, or course). It should now be clear that the examples in SAQs 7 and 8 were chosen with care! In each case, the value of $\Delta G_{\mathrm{m}}^{\ominus}$ was not completely dominant: thus the concentration (strictly, activity) ratio in the reaction quotient Q *could* be manipulated sufficiently to tip the scales in favour of the reverse reaction.